Shapi

"*'Shaping Psychology'* is a collection of revealing and stimulating interviews of leading psychologists with research interests ranging from heuristic biases to behavioural genetics. Their careers are captured in entertaining vignettes preceding in-depth discussions of their scientific contributions and their views on such contemporary issues as reproducibility, open science, artificial intelligence and the relevance of psychological research for everyday life. Recommended for those concerned about where Psychology is going in the 21st century and for students hoping to tread that fascinating path."

—Professor Trevor W. Robbins, *University of Cambridge*

"A wonderful book, full of unexpected and insightful answers to probing questions. The full sweep of psychology and the mind sciences is in full display. A must read."

—Professor Michael Gazzaniga, *University of California, Santa Barbara*

"This is a wonderful book. The biographical sketches are fascinating, and the interviews provide insight and wisdom. I highly recommend it to students, psychologists, and the general public."

—Professor "Roddy" Roediger, *Washington University, St. Louis*

"*Shaping Psychology* is an engaging and often provocative volume filled with fascinating insights from prominent researchers whose work has strongly impacted the development of psychological science. Anyone interested in the past, present, or future of psychology should read this book immediately."

—Professor Daniel L. Schacter, *Harvard University*

"The book offers important insights from leaders in the field of psychology and constitutes a very useful and original contribution to the field."

—Professor James Alcock, *York University*

"I think the project could be an important time-capsule for future generations, offering a glimpse of the psychologists and the discipline as they are found in the early 21st century."

—Professor Lisa Osbeck, *University of West Georgia and Past-President of APA Division 1, Society for General Psychology (2020–2021)*

Tomasz Witkowski

Shaping Psychology

Perspectives on Legacy, Controversy and the Future of the Field

Tomasz Witkowski
Wroclaw, Poland

ISBN 978-3-030-50002-3 ISBN 978-3-030-50003-0 (eBook)
https://doi.org/10.1007/978-3-030-50003-0

Cover illustration: agsandrew/gettyimages

This Palgrave Macmillan imprint is published by the registered company Springer Nature Switzerland AG
The registered company address is: Gewerbestrasse 11, 6330 Cham, Switzerland

Dedicated to the scientists who will perfect psychology with their bravery, knowledge, and resolve.

Contents

1
Introduction

Psychology is at present one of the most fashionable scholarly disciplines. In 2018, the American Psychological Association announced that psychology is more popular than it has ever been. It is estimated that between 1.2 and 1.6 million undergraduates take introductory psychology classes each year. According to the U.S. Bureau of Labor Statistics, overall employment for psychologists will grow by 19% between 2014 and 2024, much faster than the 7% average growth predicted for all occupations. Even the Catholic Church took notice of the field's growing popularity, quite some time ago. In his 1987 report on the state of the church, the future pope Cardinal Joseph Ratzinger expressed concerns that psychology posed a real threat to religion, that it was responsible for empty monasteries and had superseded theology (Ratzinger and Messori 1987, pp. 99–100).

Indeed, the popularity and spread of psychology is tremendous. Most illustrated magazines have advice columns edited by a psychologist to address concerns from readers. Many people have scheduled appointments with their psychologists or psychotherapists and spend a few hours every week in their offices. TV programs often feature experts (psychologists) explaining why somebody had killed, raped, defrauded money

T. Witkowski, *Shaping Psychology*,
https://doi.org/10.1007/978-3-030-50003-0_1

or committed suicide. Psychologists always show up after some major event to provide an interpretation of what happened. In most bookstores, shelves bend under the weight of volumes offering psychological support and advice. Celebrities usually discuss their psychological problems publicly and openly talk about therapeutic programs that they participated in. People socializing at gatherings exchange recommendations for psychotherapists. Psychologists often show up at crash sites or the epicenters of natural disasters. They work at schools, hospitals, hospices, social support sites and in the human resource departments of most corporations. They can be found in the police and in the army, but also in churches and prisons. Psychology truly is omnipresent.

At the same time, it is replete with controversies, flaws, uncertain claims and even myths. In recent years, the field of psychology has been rocked by numerous scandals, such as the participation of psychologists in developing methods for enhancing the interrogation of prisoners, the demasking of a tremendous scientific fraud, and multiple failures in the replication of famous experiments, primarily in social psychology. These stories have made their way onto the front pages of newspapers, and information about them has traveled well beyond the borders of the academic community. Psychology and its weaknesses are the subject of conversations everywhere, with opinions being given regardless of education and knowledge. Many authors are talking and writing about the crisis in psychology. It is not uncommon to encounter the radical opinion that psychology in general is not a science. These opinions crowd out the voices of the authorities and scientists who built its foundations. Many lay readers, students, but also psychology teachers feel lost in the flood of opinions. Despite the growing popularity of psychology, it appears to be evolving into a minefield. From time to time, one of the mines explodes, leading to a precipitous drop of societal trust in the discipline.

How strong are the foundations of our science? Are we in fact in the midst of a crisis of psychology, as the sensational headlines declare? What are our field's possibilities for the future?

While many have presented their views on the subject, credible voices answering that question are more difficult to find in the newspapers and on social media. One way to reach them is to collect the most eminent representatives of our field in one place and have them engage in a serious debate. While this is a task perhaps beyond the capacities of one individual, it is not impossible. In my view, a credible substitute for such a debate is a book containing conversations with masters of psychology. While it does not allow for a direct exchange of thoughts among them, I nevertheless believe that the careful reader, based on the words of some of our profession's most distinguished representatives, will be able to discern both the common ground they share on some issues and the distinct differences among them. In addition, compared to a traditional debate, a book has the virtue of a lifespan longer than that of other, more ephemeral forms, and after a number of years can serve as a sort of intellectual bridge linking history with the present.

In undertaking the effort of carrying out such an endeavor, I was faced with complex dilemmas. Who should be considered an "authority" in our field? Should the selection of contributors be guided by rankings, and if so, which ones? How, taking into account the limitations of space inherent to a book, can the participation of practitioners of various subdisciplines be ensured? How to avoid the important voices that sometimes fail to break through the myriad of publications preferred by the academic system of Western civilization? That very same system that doubtlessly contributed to the fact that psychology is now said to be in crisis.

In attempting to resolve these dilemmas, I understood that I would not be able to compile a list of contributors that would please everybody. However, I did everything within my power to minimize potential biases, as well as to go beyond my personal preferences. I decided to use existing academic rankings so as to ensure the participation of the most outstanding scholars and thinkers in the process of painting a picture of contemporary psychology. To this end, I primarily employed two rankings published in the form of scientific articles (Haggbloom et al. 2002; Diener et al. 2014) and one from the internet (The Best Schools 2019). The authors of the former were also at pains to apply such criteria in their rankings that would ensure the particular place occupied by a given scientist would accurately reflect that individual's contribution to psychology. The Internet ranking *The 50 Most Influential Living Psychologists in the World* was created based on the assumption that the influence of an individual can be evaluated by investigating the co-occurrence of the individual and the topic in web-accessible documents. In the creators' opinion, if a person is influential with a particular topic then this person should be often mentioned in discussions of that topic. Their approach uses machine learning and search algorithms to characterize academic influence on the web, and thereby avoids the bias of continual human intervention that infects some academic rankings.

In preparing the list of contributors, I went to great lengths for the picture painted of contemporary psychology to also encompass its foundations located in history. The discoveries made by psychology that have withstood the test of time are responsible for its scientific strength, and a detailed review of them may assist us in understanding some contemporary weaknesses. This is why, when selecting my interlocutors, I placed significant stock in the 2002 ranking titled "The 100 most eminent psychologists of the 20th century." I invited the most influential living psychologists practicing in the last century as well. This led to the inclusion of such participants as Jerome Kagan, Noam Chomsky, Michael Posner, Elizabeth Loftus, Robert Sternberg, Robert Plomin and Daniel Kahneman. The past and continuing importance of their accomplishments in forming contemporary psychology is also confirmed by the high position they enjoy in newer rankings.

Some of my interlocutors were too young to be included in the rankings of psychologists working mainly in the twentieth century, but their contribution to the development of academic psychology is so huge that they took prominent positions among the previously mentioned scientists in the academic ranking published in 2014 (Diener et al. 2014). These include neuroscientist Joseph LeDoux and social psychologist Roy Baumeister.

Using the aforementioned academic rankings in the course of creating my list of interviewees, I was to a certain degree doomed to repeat the biases present in them. One bias in the rankings of eminence is that they rely on sources that give heavy weight to Americans, or at least to English speakers. Thus, the rankings, and consequently my list of contributors, do not fully cover the entire world and give too little credit to scientists outside the USA. This should be taken into account while reading this book. Another major concern is the infrequency of ethnic minorities in my list. Although general progress has been made in terms of human and civil rights for African Americans, Asians and Latino/Hispanics, these groups are extremely underrepresented in academic rankings.

The same applies to women. The very low percentage of women in academic rankings reflects the fact that they found it difficult to be accepted to graduate programs, were virtually excluded from having professorships in universities, and usually served as research associates or assistants. And while in recent years women have dominated such fields of study as psychology, and have made deep inroads into science, it is perhaps still too soon for these advances to be reflected in academic rankings which are almost the exclusive domain of people 50 years of age and up.

Bearing in mind the fact that contemporary psychology is more than just mainstream academic research in which social, cognitive and neuropsychologists dominate, my list of contributors includes those whose activity rarely assures them a prominent place in academic rankings (or has not yet done so), but without whom the picture of our science would be incomplete. These include Brian Nosek, a leading advocate of the open science movement, which is of exceptional importance in times when psychology is experiencing a replication crisis and ways of overcoming it are being sought. It is hard to imagine an honest presentation of the condition of contemporary psychology without the involvement of advocates of the open science movement.

My invitation for an interview was extended to Scott Lilienfeld not only because he is an outstanding clinical psychologist, but also due to the fact that he applies an evidence-based approach to this area and he is a representative of a rare approach to science that can be described as subtractive epistemology. The essence of this approach consists in cleansing both scholarship and practice of false constructs that find no confirmation in empirical evidence. Engaging in such thankless "cleanup" work makes it harder to appear in academic rankings than creating a new theory, even if it later turns out to be false. Despite that, Lilienfeld is a well-known and respected scientist and skeptic, which is why I felt his perspective in the discussion on the condition of contemporary psychology may prove invaluable.

The image of mental health psychology is complemented by the words of Vikram Patel—an internationally recognized authority in the field of global mental health. His perspective is also of immense value, because it takes into account the problems of the greater part of the world's population, frequently quite different from what we focus on in our Western cultural milieu, particularly in North America. As a psychiatrist, he doesn't appear in psychologists' rankings, but his efforts have been recognized by his inclusion in the TIME 100 list of the most influential people in the world in 2015.

An even rarer sight in academic rankings are representatives of such fields as parapsychology. For years, Susan Backmore was a leading and credible scholar in the field. Having abandoned it, she can regard it critically and with distance. She is also the creator of the exceptionally interesting concept of memetics, in which she attempts to combine the biological, psychological and cultural perspectives. I felt that this synthetic and interdisciplinary kind of thinking, so rare in contemporary psychology, is also deserving of attention. Blackmore was awarded 24th place by *Best Masters in Psychology* in the list of "30 Most Influential Psychologists Working Today" (Tjentz 2013).

The voices of representatives of critical psychology—the harshest critics of mainstream psychology—are also of exceptional importance in completing our contemporary picture of the discipline. They are not to be found in academic rankings that reflect the state of a science done in a manner they systematically criticize. It would be nothing short of imprudent to ignore their voices. In science, unlike in democracy, the majority does not decide. Carefully, listening to the minority is an essential element of critical scientific thought. Erica Burnam is doubtlessly an exceptional member of this minority, who looks at psychology not only from the perspective of critical psychology, but also a feminist one, thus representing two complementary positions that are in opposition to the field's mainstream.

The list of luminary psychologists invited to speak in this book finishes with Carol Tavris who, not herself an academic psychologist, was placed on the *The 50 Most Influential Living Psychologists in the World* (The Best Schools 2019) for her profound impact on psychology. Tavris is a self-declared skeptic and feminist whose invaluable insights as a freelancer has enriched the diversity of thought in our field.

Looking at the table of contents, it should be borne in mind that its final shape is a product not only of the intentions of its author. Some of the psychologists I invited refused for various reasons to participate in my project, while others simply did not respond. I was also limited by the size of the book. At a certain point, I was forced to close the list of interlocutors, and I am painfully aware of the gaps in it. It lacks the voice of the presently strong evolutionary psychology, representatives of religious psychology and supporters of qualitative research. There are also no representatives of computational psychology, so important in the era of research on artificial intelligence and many other specialties. In closing my work on this book, I felt unsatisfied not only because of the lack of representatives of some fields, but also because of the absence of advocates of different conceptual frameworks for the science of psychology such as descriptive or phenomenological psychology.

Excusing myself by the size of the book and the fact that some representatives of the fields absent here did not respond to my invitations, I also take full responsibility for any inaccuracies in the image of modern psychology that these deficiencies may lead to. If the book is positively received, perhaps I will be given a chance to make up for these shortcomings with another volume of conversations.

Taking into account the criteria and methods for selecting my interlocutors, it should be kept in mind that this book does not aspire to be a comprehensive, objective scientific interrogation and should under no circumstances be treated in this way. However, I am hopeful that, in this selection of voices with its inherent limits, readers will find multiple intriguing reflections, sources of inspiration and topics for discussion, and that as a whole it will help to develop a fuller picture of psychology than the fragmentary image portrayed in the media. In expressing these hopes, I also owe the reader a few sentences on the form of the book. I chose interviews because I believe that people have a need for direct contact with authorities. Only a select few enjoy the privilege of participating in lectures and seminars led by the interviewees collected in this volume. An even smaller number has the chance to speak with them and ask questions. Encountering all of them in one place, even at the most prestigious conference, is simply not possible. People generally come into contact with their scholarship and thought by way of books and articles, which, in line with publication requirements, are devoid of any personal reflections. This contact is even more frequently achieved through textbooks, which summarize and present only the results of their work. There are no reflections on the future of the field, no advice as to following one's own career path, nor answers to the questions and issues raised in the media. The need for direct contact with authorities means that the most popular forms are those which can substitute for such contact, like TED Talks or interviews published on YouTube. Another form of contact that remains consistently popular are television, radio and print media interviews. Yet rarely do viewers, listeners and readers have the opportunity to experience interviews with exceptional individuals conducted by one person asking them similar questions. I hope that the present volume meets readers' need for direct contact with the knowledge and perspectives of leading authorities in the field of psychology.

In conducting the interviews, I decided to allow my interviewees the greatest possible freedom to speak, attempting to direct the conversation towards subjects I find interesting rather than to probe or challenge them in a systematic way. This form of conversation is a product of my humility in the face of the knowledge and experience enjoyed by my interlocutors. I also do not think that one individual would be capable of a thorough and credible exploration of all areas of psychology represented by the contributors. Despite the free tenor of the conversations, I attempted to ask all of my contributors' questions about:

- studies, experience and achievements of the interviewees in their areas of specialization;
- the root causes of problems affecting the field of psychology and comprising what is being referred to with increasing frequency as the crisis in psychology;
- ways these problems can be resolved;
- achievements of psychology as a scientific discipline;
- the challenges facing psychology; and
- recommendations for people just getting their professional career in psychology started.

The order in which these issues are addressed in the interviews is not fixed, but rather results from the natural character of each conversation. The questions have not been standardized, in order to avoid becoming overly schematic and preventing boredom in the reader. Answers to some of these questions are not to be found in the transcriptions of the conversations. This is because some of the contributors did not consider themselves competent to respond to those questions, or simply preferred not to. In these cases, I have spared the reader's time and patience by removing sterile passages from the transcripts, with the consent of the interviewees.

All the conversations are preceded by a short profile of the interviewee and his or her research, which should contribute to a better understanding of the issues raised. Each interview is accompanied by a *References* section with titles referred to by both the author and his interviewee, as well as *Selected Readings* containing some of the most significant publications of the interviewee, mainly books.

Because the order of presentation of outstanding figures in any field bears with it the suggestion of an implicit hierarchy, in the book I have decided to present the interviews in the order in which they were conducted. The book's table of contents, particularly the order in which the conversations are presented, should in no event be considered a ranking of my contributors.

References

Diener, E., Oishi, S., & Park, J. (2014). An incomplete list of eminent psychologists of the modern era. *Archives of Scientific Psychology, 2*(1), 20–31.

Haggbloom, S. J., Warnick, R., Warnick, J. E., Jones, V. K., Yarbrough, G. L., Russell, T. M., & Monte, E. (2002). The 100 most eminent psychologists of the 20th century. *Review of General Psychology, 6*(2), 139–152.

Ratzinger, J. C., & Messori, V. (1987). *The Ratzinger report*. San Francisco: Ignatius Press.

The Best Schools. (2019, June 14). *The 50 most influential living psychologists in the world.* Retrived from https://thebestschools.org/features/most-influe ntial-psychologists-world/.

Tjentz. (2013, September 5). 30 Most influential psychologists working today. *Best Masters in Psychology.* Retrived from https://www.bestmastersinpsych ology.com/30-most-influential-psychologists-working-today/.

2

Elizabeth F. Loftus: Cognitive Psychology, Witness Testimony and Human Memory

Zealous conviction is a dangerous substitute for an open mind.

Elizabeth F. Loftus

T. Witkowski, *Shaping Psychology*,
https://doi.org/10.1007/978-3-030-50003-0_2

Filmmakers rarely make blockbusters out of the lives of scholars. That said, the first director who decides to make a film about the life and times of Elizabeth Loftus won't have to do very much to keep viewers' attention and build the suspense. Her biography is a ready-made script, full of dramatic plot twists, a riveting struggle of good against evil, honor against dishonor, and truth against lies. I am convinced that sooner or later we will see the history of this exceptional life on the big screen. I arrived at this belief in the course of reading pages and pages of biographical material, interviews and recollections prior to my conversation with Loftus. But the events that make her life's history an attractive film subject are, to the person who experienced them, obstacles that absorb a tremendous amount of energy to overcome. Only a very few can successfully manage them, and fewer still rise above them while remaining faithful to ideals. Among these few, we may invariably find Elizabeth Loftus.

Beth Fishman, the girl who would become Elizabeth Loftus, early on in her childhood was put to the test in a way that would break many. When she was 6, a babysitter molested her. When Beth was 14, her mother drowned in a swimming pool. The obituary called it an accident, but Beth's father suspected suicide. Two years after she lost her mother, Beth lost her home. A brush fire destroyed her house together with over 400 other homes in broader neighborhood. Despite these difficult experiences in 1966, and despite being raised to expect little more from life than being a wife and mother, she entered Stanford's graduate program in mathematical psychology. She was the only woman admitted to the program that year, and her classmates wagered among themselves whether she would graduate.

But she did and she started her research focused on the organization of semantic information in long-term memory. But that what she was doing was not something she wanted to devote her life to. She decided to seek out research fields of greater social relevance and begin a new line of research into how memory works in real-world settings, beginning the empirical study of eyewitness testimony. Soon she developed the misinformation effect paradigm, which demonstrated that the memories of eyewitnesses are altered after being exposed to incorrect information about an event—through leading questions or other forms of post-event information; and that memory is highly malleable and open to suggestion. The misinformation effect became one of the most influential and widely known effects in psychology, and Loftus' early work on misinformation generated hundreds of follow-up studies.

Loftus, however, was not only interested in laboratory work. She was curious about how her discoveries applied to real-life situations, in real court cases, to real witnesses. So she asked for permission to observe courtroom trials. One of them led her to write an article titled "Reconstructing Memory: The Incredible Eyewitness," which was published in the popular science magazine *Psychology Today* (Loftus 1974). To her surprise, this seemingly insignificant piece led to a flood of phone calls from lawyers requesting her help with their cases. This was the start of a new chapter in her life, which led to her participation as a memory expert in over 250 hearings and trials. She consulted or testified in dozens of famous cases: Ted Bundy, O.J. Simpson, Rodney King, Oliver North, Martha Stewart, Lewis Libby, Michael Jackson, the Menendez brothers, the Oklahoma City bombing, and many more.

After Loftus had become a bit bored with the routine of the standard eyewitness cases, she was asked, in 1990, to participate in the unusual case of George Franklin, who stood accused of murder; but the only evidence against him was provided by his daughter, Eileen Franklin-Lipsker, who claimed that she had initially repressed the memory of him raping and murdering her childhood friend, Susan Nason, 20 years earlier, and had only recently recovered it while undergoing therapy. Loftus took an interest in the case because while she gave evidence about the malleability of memory, she had to concede that the research on memory distortion involved changing memories for small details of an event. This was somewhat different from the particular kind of memory Franklin-Lipsker was claiming to have, namely witnessing a rape and murder, and enduring years of other traumas that had supposedly been repressed. Could these huge memories be planted? Loftus was not aware that participation in this case would not only radically change her research, but would also turn her entire private life upside-down.

Admitting that she did not know whether it was possible to implant false memories for entire events that had never taken place, Loftus began work to find out whether some of these recovered memories might in fact be false memories, created by the suggestive techniques used by some therapists at the time. After many attempts, she developed together with her students Jim Coan and Jacqueline Pickrell the "lost in the mall" technique. The method involves attempting to implant a false memory of being lost in a shopping mall as a child and testing whether discussing a false event could produce a "memory" of an event that never happened. In her initial study, Loftus found that 25% of subjects came to develop a "memory," for the event which had never actually taken place (Loftus 1999). She would later call these "rich false memories." She thus proved something novel and powerful about the malleability of memory.

This was the beginning of the memory wars. Even before the article describing the "lost in the mall" study made it to press, it was met with the intense criticism of supporters of repressed memory therapy (Loftus 1999), whose interests she directly threatened. For Loftus, this was the beginning of an exceptionally difficult period of hate mail, death threats, public attacks, and ostracism, which we spoke about during our interview. But the worst was yet to come. In 1997, David Corwin and his colleague Erna Olafson published a case study of "Jane Doe" (real name Nicole Taus), which was, in their opinion, an apparently bona fide case of an accurate, recovered memory of childhood sexual abuse. Skeptical, Loftus and her colleague Melvin Guyer decided to investigate further. Using public records and interviewing people connected to Taus, they uncovered information Corwin had not included in his original article—information that they thought strongly suggested Taus' memory of abuse was probably false. While Loftus and Guyer were conducting their investigation, Taus contacted the University of Washington and accused Loftus of breaching her privacy. In response, the university confiscated Loftus' files and put her under investigation for 21 months, forbidding her to share her findings in the meantime. It took Loftus two hard years to win a letter of exoneration and another six years to get rid of Jane's subsequent lawsuit, which went all the way to the California Supreme Court. In their report on Jane Doe, published in the *Skeptical Inquirer* in 2002, Loftus and Guyer affirmed their duty to uncover "the whole truth" and presented the results of their investigation (Loftus and Guyer 2002a, b).

Although eventually exonerated of any wrongdoing, this was the straw that broke the camel's back for Loftus. She could not forgive the University of Washington for the manner in which they had handled the most difficult case she had ever encountered. She moved to the University of California, Irvine.

Today, Loftus is most interested in what some term "memory engineering." Is it possible to insert false memories that can bring positive effects? This is the main question she is seeking the answers to, and much as in the case of most other questions she has taken on, her efforts are proving effective. For example, her more recent research demonstrated that we can convince people of their having been averse to certain foods in their childhood, and they will begin to avoid those foods if we implant the memory well enough. While this example may seem a bit frivolous, the fact that false memories can modify our present behaviors is another of Loftus' fundamental discoveries, which creates limitless possibilities for developing new therapeutic approaches. Apart from the possibilities that her present research is developing, there are also numerous ethical questions.

Professor Loftus, most lay people imagine that a psychologist's work consists mainly in examining people using tests and questionnaires, conversing with them, and interpreting their responses. Yet there was a time in your professional life when you were accompanied at lectures by plainclothes bodyguards, and you yourself learned how to shoot. When I talk about this, people wonder what, exactly, a psychologist had to do to fear for her life in public places.

The trouble that I faced started after I began questioning some of the practices of some psychotherapists. What was happening was that people were going into therapy with one problem—maybe they were depressed, maybe they had an eating disorder—and they were coming out of this therapy with another problem, a different problem. Horrible childhood memories of horrible abuse, allegedly perpetrated upon them by their parents or other relatives, or former neighbors. And when I began to investigate these cases, it appeared as if it was some of these psychological practices that were leading people to develop false memories. And when I began to write about this, it made a lot of people mad. It made some of the therapists mad. It made some of the patients who thought they'd recovered these memories mad. It made some lawyers, who wanted to sue on behalf of these accusing patients, mad. For a while, there were these threats. Now things have died down a little bit but the problem is not over.

How long did this period of hatred last?

I first started learning about these cases around 1990, and then I co-authored a book called *The Myth of Repressed Memory* about this problem in 1994. I had already published a big article in *American Psychologist* in 1993, so probably that article and that book brought me to the attention of many of the people who would be angry about these ideas. So, throughout the 90s this was a problem. What happened in the mid and late 90s is that people who once thought they had recovered repressed memories of horrible abuse began to realize their memories were false, and they then sued their former therapists for planting false memories. And that generated the tide of change, because millions of dollars were then paid out to patients who had been led down this horrible path.

And has this period definitely concluded now?

No, there are still cases of this. Things are a little bit different now—for example, the clergy abuse cases. You have some genuine victims of abuse by priests and other religious figures. There's nothing really fishy about the memories, they aren't claiming they repressed them, but when they go public, it brings hundreds of other people, not all of whom were abused. But some of them claim to have recovered repressed memories, and they try to use the initial accusers as corroboration for their own story. So, there are still problems out there. And there are still families that are getting destroyed by these kinds of dubious accusations.

Of course, there is no justification for the manifestations of hatred towards you by people whose interests you threatened, but we can try to understand them or rationalize their behavior. However, you also had similar experiences at the hands of scientists—people that we're trained to think hold the truth up as one of the highest values. Why did these people in particular attack you?

Their social and political beliefs and opinions were just so strong that they wanted to ignore the science. They couldn't help themselves, and that's how sometimes you would see scientists get into this controversy and insist that the experiments that I and others did weren't relevant, or that we were ignoring important data. Every now and then they would say things like "science isn't the only way to know things."

During that worst period, did you ever think about walking away from it all?

I don't think so. One of the worst things that happened was when one of these individuals, whom we'll call Jane Doe, came to believe that her mother had sexually abused her when she was a child. I believe this was because of the suggestive things she was put through. She filed a lawsuit that we had to defend against for many years. This went on until 2009, in fact, but finally that case was resolved.

Was that the most difficult moment for you?

Actually, before Jane Doe sued me and my co-author and the magazine, she had complained to my former university that I was looking into her life and she was upset. My former university then began an investigation of me, and that was a bad period, because I had no idea how it would end up. Potentially my job was in jeopardy, but eventually after a couple of years of investigation I was exonerated of any wrongdoing and could get back to my work.

In past interviews, when speaking about yourself you have emphasized how important those moments in life were when you realized that you wanted to do research of greater social relevance. It's fair to say that not only did you choose a research field of greater social relevance, but that you also touched a nerve. How many researchers-psychologists think like you do and choose a similar path?

When I went from doing very theoretical memory work and began to study the memory of witnesses to crimes, accidents and other legally important events, that was in the early 1970s. Not very many people were doing that kind of work. While most other memory scientists were doing very theoretical work with very simple stimuli, I started showing people films of accidents and crimes, and studying the memory of these much more complex events. Today, lots of people do that kind of work.

Are those who prefer to remain in the world of abstract relations discouraged from engaging real issues by the potential consequences they might experience?

It's possible that some are. But it's also the case that for a long time within the field of psychology, if you did very theoretical kind, abstract work, that was held in higher regard than if you did work that had obvious practical applications. Today I think people and many of the funding agencies do want you to think about how this will fix some mental health or social problem. Maybe it discourages some people from tackling sensitive topics like child abuse.

When discussing research with greater social relevance, we hit on another important problem that is afflicting contemporary psychology. I mean the shift from direct observation of behavior, widely regarded as an advancement in the development of scientific methodology, to introspection. This was demonstrated aptly in a 2007 article by Baumeister and collaborators, and recently this year was confirmed by Polish scientist (Doliński 2018). What are your views on the issue?

I consider myself a cognitive psychologist. Sometimes I am referred to as a social psychologist, but actually I'm a cognitive psychologist, I was trained in cognitive psychology, my academic heroes are cognitive psychologists, the journals I typically read are cognitive psychology journals as opposed to the social psychology journals. So, in my field, the field of memory, we are looking at behavior. In the experiments I do, there is typically some ground truth. There was an event, and then you can look at the behavior and ask how concordant and accurate people are when they try to remember those events. So, we're looking at behavior all the time. Maybe in social psychology, where there's more focus on attitudes or some other issues, there is much more self-report going on. But we know that self-report can be contaminated because people want to paint themselves in a good light and they might exaggerate their experiences and responses to look better. So you do want to look at behavior. Maybe some combination of self-report and behavioral studies is the right combination to teach us about the world.

The shift from behavior to introspection is not the only one trouble our field is undergoing. Recent revelations of scholarly fraud, the absence of representativity in psychological studies, and problems with replicating the results of numerous experiments have led people to speak openly of a crisis in psychology as a science. Yet many

scientists deny this is the case. What is your opinion—are we really in the midst of a crisis, and if we are, what are its root causes?

It's certainly useful to ask the question of whether there are publication pressures: are the journals wanting to publish something that is brand-new and seems exciting, maybe even counterintuitive, but into which maybe not enough work has gone into? So, yes, every now and then some study gets published and a phenomenon gets reported, and then it turns out it's tough or not possible to replicate. But I also think that science corrects itself. People will eventually figure out when something isn't replicating, and they'll communicate that. I have seen some poor attempts at replication, where someone calls out the original study when their replication wasn't even a good effort to replicate, and they impugn the integrity of the original investigator. I don't think that's very good for morale and good will. We can do better in terms of being sure that our findings are likely to replicate, but we already had some ways in which we could accomplish that before we had all these scientific police officers roaming the streets.

Could you explain, what do you exactly mean while talking about scientific police officers?

There are individuals who are scrutinizing the work of scientists. Some of them have great intentions, and want to correct some long standing practices that were possibly contributing to flimsy phenomena being published. Criticizing people for using more liberal statistical tests than they should have could be seen as constructive criticism. But some of these scrutinizers are act more like hostile bullies and attack practices that are not so clearly wrong. Or they implicitly accuse an investigator of fraud, when the "mistake" was a much more innocent one.

Do you think that we as scientists should do something to overcome crisis in psychology? And if so, what are the most urgent tasks that psychologists are faced with today?

Some things are being suggested, like reporting a pre-registered hypothesis in advance, and stating in advance how many subjects you plan to run. One of the prescriptions that I don't like is the idea that you cannot peek at your data. It's so tempting when an experiment is underway and you're curious. You want to know how it's going, and you don't want to wait for three months to get the answer when you're halfway there—so you can get a hint, while you're excited and interested. It's still tempting to want to peek, but I think as long as you tell people what you've done and are open about it, there are some steps toward fixing the problem. If there is a problem.

Could we say a consequence of all these methodological problems is a decline in trust towards our field? What are your views on the matter?

I don't know. When you get publicity about supposedly one hundred studies while only 40 replicated and the other 60 did not, people will wonder about psychology and about social psychology in particular, because those seem to be the ones that people have been trying to replicate. But when you really look into the replications were not always a fair attempt.

Low replicability of research is not the only one reason of the decrease in trust in our field. From time to time, psychologists engage in ethically questionable activities just as they did at the beginning of the twenty-first century, when psychologists affiliated with the APA were involved in work on techniques for interrogations to be applied as part of the war on terror. This is not the first time that people from our discipline took active part in researching and developing methods used against other people. We can recall the example of the CIA-inspired MKUltra program. Looking at it from a certain perspective, this is also research "of greater social relevance," but tremendously different from what you do. How do you judge the participation of psychologists in perfecting torture techniques, which is probably a fair phrase to use in referring to these "interrogation methods"?

My co-authors and I recently published a study showing that sleep deprivation leads people to be much more likely to confess to a wrongdoing that they did not do when compared to people who were not sleep deprived. I think this is an important study because it tells us a little something about sleep deprivation, which is one of the ingredients in torture methods. And I think it's our skills as psychologists and our methods that allow us to learn something about at least this one element. It's an important contribution to science, and I think it says something about the policy, namely, that this is a policy that might not produce very good outcome. So, I certainly see a role for psychologists in doing work that will educate us about the effects of these torture elements and their outcomes.

But this is a little different story, because you are talking about neutral results which can be used in a good way or in a bad way.

Yes.

I mean the case of the psychologists who were employed by the CIA and were actively researching and looking for better methods of interrogation. They worked out methods which were not useful in gathering better information, but which caused much pain for the interrogated people.

I'm not a member of the American Psychological Association, but apparently that organization has rules against psychologists participating in efforts to figure out how to torture people better. So, I think you need to ask somebody who is more of a clinician and who is a part of the American Psychological Association community.

After these reflections on the condition of contemporary psychology, let's return to your history for a moment. I read a statement in which you said that at the beginning of your career you were fascinated by the approach of B.F. Skinner. Could you say a little more about this?

Yes, I was an undergraduate student at UCLA. And when I was at UCLA, I began taking courses in psychology. Even though I was majoring in mathematics, I ended up taking so many elective courses in psychology that I finished with degrees in both mathematics and psychology. During my studies in psychology at UCLA I learned about Skinner. It was the idea of patterns and reinforcement and the elegance of his work that excited me. Being given a rat and being involved in trying to train this rat to press a lever. Ultimately, after I received my Ph.D., I actually had a chance to meet Skinner and had a number of lunches with him.

How do you remember him?

I was spending a year at Harvard, and I sent a letter to Skinner—this was, of course, way before the internet and email. I wrote that I was an experimental psychologist and that I had held my Ph.D. for five years, and I was at Harvard for a year. I said, "nothing would give me greater pleasure than to have lunch with you once this year." And the next thing that happened is my office phone rang, and he called me and said "Hi, it's Fred Skinner." And I was shocked! We had lunch, and he talked the whole time about the second volume of his multivolume autobiography. At the time he was working on this volume that covered from age 9 to 20, the early years of his life. He was telling me about his writing and he didn't really ask me very much about me, but at the end of the lunch he said "You're such a fascinating conversationalist, would you like to have lunch again?" So, we began a series of lunches at which I would get to listen to his ideas, and I mostly just listened because he didn't ask very many questions of me.

Wasn't this inspiring?

Well, after a while, when somebody doesn't turn to you and say "Tell me about you," it's not that much fun.

Aha. During your decades of research, have there been other psychologists who had a similarly strong influence on you?

Certainly. When I was in graduate school at Stanford, I was doing work on Computer Assisted Instruction, and I took a course with a psychology professor named Jonathan Freedman. He was wanting to do some studies on memory, and he asked me if I wanted to join him and participate in those studies. That was how I began to study memory—the theoretical, semantic memory, not the kind of eye-witness memory that I would ultimately do later. So, he's certainly responsible for getting me interested in memory as a topic and teaching me how to be an experimental psychologist: how to design a study and produce the materials, run the subjects, analyze the data, and write the manuscript. I owe a lot to him.

Who would you name as a model to follow for young people just getting their careers off the ground?

One piece of advice I've often given to people who are thinking about going to graduate school is that if they think they can, and if they think they want an academic career, they should look up the productivity. And certainly find somebody who's working on a topic they're interested in, but you want people who are publishing with students, because students will need those publications to secure an academic job. So, I think that's important. Once you start your academic job, hopefully you have an idea of who you are and what kind of studies you want to do. Again, it is important to publish early in one's career, because the tenure decision is going to be based on that. At least that's some advice for people who aspire to a career in academia.

I think we can say without an ounce of exaggeration that you proved to the world that memory does not function like an audio cassette or film reel we can simply rewind to the desired moment and start watching, that memory can also include events which never actually took place, that much can be erased from it, and what's left behind is often incomplete and inaccurate. Your discoveries have helped many people. At the end of the 1990s, you left behind the study of distortions of memory in favor of a sort of memory engineering—the implantation of memories in order to achieve specific effects. Was this decision the result of your belief that the problems that inspired your research had been solved?

There was a period where I was with my then post-doc, Daniel Bernstein, and my two then-graduate students who had come with me to the University of California when I moved here. We took an interest in the repercussions of having a false memory. If I plant a false memory in you, this will affect your later thoughts, or your later intentions, or even your later behavior. We talked a lot about finding a way to study that, and finally we decided that what we would try to do is to plant a false memory of getting sick eating a particular food. When we succeeded in planting that false memory, we found that people weren't that interested in eating those foods after they had developed the false belief or memory. So, if you convince someone that they got sick as a child eating strawberry ice cream, they're not as interested in having strawberry ice cream. And this was our first clue that you could plant these memories and it could affect behavior that occurred often a bit later. So, we did a lot of studies of that sort, trying to expand our knowledge of false memories and their applications.

Do you have any concerns about ethical issues that can arise out of your studies on what some term memory engineering?

Oh yes, certainly, when I talk about this I do. Because we know how to plant false memories and we know it can affect people's behavior—it can make them less interested in eating fattening food, more interested in eating healthy food, less interested in the particular alcohol they drink—it does raise the question of whether we can ever affirmatively use these techniques on people, maybe to allow them to live a happier or healthier life, or should we ban their use. People sometimes cringe at the idea that other people out there might want to plant false memories in them and affect what happens in the course of their lives, but I also suggest that, for example, if you could plant a false memory that would make people avoid a fattening food that would make them less likely to become overweight or obese, less likely to develop diabetes, less likely to develop heart trouble, less likely to have a short lifespan, then maybe they would be better off. And there are some times when there's some kind of a trade-off between truth and happiness.

Your experiments and publications are among the highest-rated in terms of citations and in lists of popular studies, and you have received multiple awards. What I am most interested in (and I hope our readers are too) is this question: which of your discoveries you feel are the most important in the development of science, and why?

I don't think that you can really point to one article or one scientific study. It's not like an author can say "here's my book 'War and Peace,' this is my great contribution." The thing I'm most proud of is the body of work that has revealed so much about the malleable nature of memory, about the fact that it's not just a recording device, but that it's a constructive process. And that it does probably contain bits and pieces of fiction mixed in with facts. I think what's important is the body of scientific work that reveals this truth about memory.

Readers interested in psychology can read quite a lot about your previous accomplishments, but the majority of us are undoubtedly interested in the issues you are presently engaging. What questions are you trying to find answers to these days?

In the last couple of years we have become interested in something we call "memory blindness." In these memory blindness studies you have somebody who gives you a report about a memory they have. They tell you, for example, that a thief who stole a wallet was wearing a green jacket, and some time later you come back to them and you feed them information about their earlier response, but it's wrong. You tell them, for example, that they had said the jacket the thief wore was brown, then ask if they remember what brand that jacket was. So, we're now telling them they gave us a response that was different from the one they previously gave. Do people even detect it? Often, they don't, and especially when they don't detect it they will often succumb to the suggestive influence, distorting their memory in the direction of that mistaken information. So, this is leading us to all kinds of questions: when do people detect that something is wrong? When do they notice? How and when can they tell you or alert you if they noticed that they've detected that something is wrong? What surprises us is that people often don't detect. And then they're influenced. So, a number of the graduate students in my lab have been doing studies of what we call "memory blindness." This is just one example.

What are other issues we have not talked about yet but you would like to mention to our readers? Such as a message about contemporary psychology, a big issue or important question about psychology?

One other thing in our field we're interested in is how your personal biases, either your political biases or your social biases, can make you more susceptible to receiving misleading information that is consistent with those biases. And there are number of recent studies we're involved in which demonstrate that, indeed, people are more inclined to accept suggestive information when it fits with what they already feel.

Do you think this is a problem which is strictly connected with recent political changes and issues with our contemporary political life where everybody has access to social media, or this is a more universal problem?

This is a universal problem, but it also applies to the political context. People's political beliefs set them up to be especially susceptible to being contaminated with false information that fits with those beliefs. This is something that's now being increasingly documented, but what we end up doing about it is something the next generation of psychologists will have to figure out. But I think it doesn't bode well for the problem of fake news, because it's been shown that even when you have a sense that it's fake, you can still be influenced by it.

Professor Loftus, I believe that your life is a ready-made script for a feature film that I hope I'll get to watch some day. In thanking you

for our conversation, I would also like to express my wish that no more sudden plot twists get thrown into the script of your life.

Selected Readings

Garry, M., & Hayne, H. (Eds.). (2007). *Do justice and let the sky fall: Elizabeth F. Loftus and her contributions to science, law, and academic freedom*. Mahwah, NJ: Lawrence Erlbaum Associates.

Loftus, E. F. (1975). Leading questions and the eyewitness report. *Cognitive Psychology, 7*, 560–572.

Loftus, E. F., Doyle, J. M., & Dysert, J. (2008). *Eyewitness testimony: Civil & criminal* (4th ed.). Charlottesville, VA: Lexis Law Publishing.

Loftus, E. F., & Hoffman, H. G. (1989). Misinformation and memory: The creation of memory. *Journal of Experimental Psychology: General, 118*, 100–104.

Loftus, E. F., & Ketcham, K. (1991). *Witness for the defense; the accused, the eyewitness, and the expert who puts memory on trial*. New York, NY: St. Martin's Press.

Loftus, E. F., & Ketcham, K. (1994). *The myth of repressed memory*. New York, NY: St. Martin's Press.

Loftus, G. R., & Loftus, E. F. (1976). *Human memory: The processing of information*. Hillsdale, NJ: Erlbaum Associates.

References

Baumeister, R. F., Vohs, K. D., & Funder, D. C. (2007). Psychology as the science of self-reports and finger movements: Whatever happened to actual behavior? *Perspectives on Psychological Science, 2*(4), 396–403.

Corwin, D., & Olafson, E. (1997). Videotaped discovery of a reportedly unrecallable memory of child sexual abuse: Comparison with a childhood interview videotaped 11 years before. *Child Maltreatment, 2*(2), 91–112.

Doliński, D. (2018). Is psychology still a science of behaviour? *Social Psychological Bulletin, 13*(2). Retrieved from https://doi.org/10.5964/spb.v13i2.25025.

Loftus, E. F. (1974). Reconstructing memory: The incredible eyewitness. *Psychology Today, 8,* 116–119.
Loftus, E. F. (1993). The reality of repressed memories. *American Psychologist, 48*(5), 518–537.
Loftus, E. F. (1999). Lost in the mall: Misrepresentations and misunderstandings. *Ethics and Behavior, 9*(1), 51–60.
Loftus, E. F., & Guyer, M. (2002a). Who abused Jane Doe? The hazards of the single case history, Part I. *Skeptical Inquirer, 26*(3), 24–32.
Loftus, E. F., & Guyer, M. (2002b). Who abused Jane Doe? Part II. *Skeptical Inquirer, 26*(4), 37–40, 44.
Loftus, E. F., & Ketcham, K. (1994). *The myth of repressed memory.* New York, NY: St. Martin's Press.

3

Jerome Kagan: Temperament, Developmental Psychology and Methodology

Asking the right question is more important than trying to have a publishable paper.

Jerome Kagan

T. Witkowski, *Shaping Psychology*,
https://doi.org/10.1007/978-3-030-50003-0_3

In summarizing his life and work, Jerome Kagan goes against the grain and points to six entirely unrelated events as the sources of his success. In doing so, he overlooks the very important fact that the coincidences (Kagan 2007) he cites were subjected to his exceptionally inquisitive mind, remarkable persistence and passion for learning and discovery. An ordinary person would likely have failed to follow their path to the place where I encountered Kagan in the course of our conversation—a position that is among the most renowned living psychologist. This, at least, is what we can glean from the ranking of the 100 most influential psychologists of the twentieth century, published in 2002 in *Review of General Psychology* (Haggbloom et al. 2002). Jerome Kagan is to be found there in 22nd place, even above Carl Jung (23rd), the founder of analytical psychology, and Ivan Pavlov (24th), who discovered the reflex bearing his name.

He is probably best known for his experiments on temperament, which he describes during our conversation. In fact, in 2004 *The Boston Globe* even gave him the nickname "The Temperamentalist" for being the scientist who restored legitimacy to the ancient notion of temperament (Shea 2004). However, he made his discoveries in a time of near-universal belief that the individual's environment is the primary determinant of their psyche. This understanding of development manifested itself with particular clarity in John Bowlby's attachment theory which claimed that the bond between mother and infant, as measured in the first year, plays a key role in later emotional and even intellectual growth. Kagan was among the harshest critics of attachment theory. In spite of the quite strong empirical evidence against attachment theory, it retains the loyalty of a small group of contemporary psychologists, which, in Kagan's view, results from the fact that many scientific conceptions are grounded not in the results of experiments, but rather in the life experiences of their creators and their culture.

Attachment theorists are not the only ones whose claims have run up against Kagan's fierce objection. Supporters of evolutionary psychology (who see Darwinian selection at work in many human psychological traits) have also come in for criticism at his hands. Particularly intense fire was trained on Judith Rich Harris, author of "The Nurture Assumption" (1998), who argues that parents play little role in shaping their children's basic personality traits. After her book was published in 1998, he said during an interview with *Newsweek* "I am embarrassed for psychology," and in the *Boston Globe,* he commented on Harris' book by simply saying "It had nothing to do with science. If the media had not hyped it - partly because it was so crazy - nobody would know about her."

For some scientists, his rejection of attachment theory is in conflict with his own criticism of Harris. Indeed, in this criticism Kagan emphasizes the importance of environment. In reality, his approach emanates from his conviction of the interactive nature of those two powerful forces—genetics and environment.

Jerome Kagan is an uncompromising scientist who does not hesitate to oppose the majority when he feels they are attempting to impose their views. On several occasions, he has sparked outrage among public opinion and the media. One such statement came in 2012 during an interview for *Der Spiegel*, when he responded to the interviewer's question of whether ADHD is just an invention. "That's correct; it is an invention. Every child who's not doing well in school is sent to see a pediatrician, and the pediatrician says: 'It's ADHD; here's Ritalin' (Kagan 2012b). In fact, 90% of these 5.4 million kids don't have an abnormal dopamine metabolism. The problem is, if a drug is available to doctors, they'll make the corresponding diagnosis." He found himself tossed almost immediately into the eye of a storm that had erupted around him, and his words were repeated on countless occasions. This comes as no surprise, as they referred to over 5 million American children who had been given such a diagnosis; to their parents as well, for whom the diagnosis entailed certain consequences; to teachers; to the doctors who had made the diagnosis, and generally followed up by prescribing medication to their patients, most frequently Ritalin. And if we add to this both the pharmaceutical companies producing the medicine and the pharmacies selling it, we have a group as numerous as the population of the state of New York. Yet we also know that ADHD is eagerly diagnosed outside the USA as well.

Jerome Kagan can be described using his own sobriquet: "Strong, silent, Clint Eastwood type," which would explain why he handles intellectual jousting quite well, and does not allow himself to be affected by the labels slapped on him: iconoclast, extremist, dogmatist, etc. In my view, however, they are detrimental to many of the fundamental postulates he advances, and are a defense mechanism guarding against uncomfortable statements. For years, Kagan has criticized psychologists' passion for abstraction, attacking researchers for conducting experiments that ignore the context, and for failing to take into account the expectations of study participants placed in experimental situations. He has warned against relying only on a person's verbal reports of their traits or past experiences as the basis for bold conclusions about behavior or mood; he also points out a plethora of other weaknesses of our field. The weight of his criticism cannot be overstated, but nevertheless, when someone assigns to the author of such warnings the labels of extremist or dogmatist, it becomes much easier for us to deprecate them.

Professor Kagan, during one of your lectures you described a dream in which you meet people working on the renovation of an eighteenth-century house, and you offer to help them out. In response, one of them asks you to work on the restoration of a beautiful but run-down chest. When the chest is ready, you bring it to the renovated house, but the house isn't there. During the time you were working on the chest, the house had fallen apart. This is essentially a very bitter metaphor of your professional career. Are things in psychology really so bad that what we have today is worth far less than what you encountered when beginning your career as a scholar/researcher?

The state of psychological research in the domains that study humans has not advanced as rapidly as many hoped. There are several reasons for this sad fact. One is that government support of basic research on behavior, cognition or emotion that does not involve genes or brain measures has become hard to obtain. It is difficult, though not impossible, to make an important discovery if the investigator has limited funds. Second, private foundations have become reluctant to fund basic research and prefer to support work that is likely to have a benevolent effect on humans or society. Third, the number of non-tenured faculty in psychology departments has increased by a large amount. This means that many junior faculty are competing for a small number of professorial appointments. Because deans and chairpersons use number of publications in prestige journals as the primary bases for promotion, rather than quality of the work or teaching talent, junior faculty strive to have many published papers over a short interval. This pressure to publish is not conducive to brilliant insights.

Among well-known psychologists, you are probably the harshest critic of our field. In your books, you highlight both misconceptions that have taken root over generations, as well as methodological problems in psychology. You also offer proposals for resolving them. Your unquestioned position in the profession makes it difficult for others to overlook your admonishments. Do you think that your efforts have led to changes of at least some beliefs and bad practices among psychologists?

There have been some positive changes in the practices of psychologists who study humans but I cannot be sure they are due to my critiques. These include publishing scatter plots, interest in human temperaments, and an acknowledgment by investigators who use questionnaires that their conclusions are limited to this source of evidence.

Not much. In your view, which of the misconceptions and bad practices are the most deeply rooted, the most difficult to change, and why?

The kinds of practices and habits that should be changed involve questioning the practice of: (1) the increased use of questionnaires as proxies for behavior, (2) an indifference to the context of observation and the source of evidence for an inference, (3) a preference to affirm a priori hypotheses by searching for a relation between single predictors and single outcomes, rather than beginning with a puzzling phenomenon and probing its features and consequences with an open mind, (4), assuming that a popular word, such as anxiety, loneliness, fear or learning, names a single natural kind that can be observed with a brain measure or behavior, (5) the reluctance to abandon DSM categories as outcomes or predictors, even though every category is heterogeneous in its etiology, (6) using covariance to control for the contribution of social class or gender despite the warnings by eminent statisticians (e.g., Helene Kraemer of Stanford) that this practice often produces invalid inferences.

That's quite a lot. Have the practices you list contributed to what is referred to with increasing frequency as the crisis in psychology? What I have in mind is primarily the problems with replicability of

investigations, academic fraud, low statistical power of psychological research, decline effects, etc.

Yes.

What should be done to change this?

I do not know what can be done to alter these habits. I suspect it will require more investigators demonstrating the validity of the above critiques.

Do you think that the initiatives proposed and undertaken by open science movement can help to solve some of these problems?

I don't know.

How does the professional community react to your disruptive activities?

As of 2018, American psychologists have responded to my books with silence. I have received neither hostile nor praiseworthy messages.

Have reactions by psychologists from other parts of the world been more enthusiastic?

European psychologists have been more receptive than Americans.

Considering that your books address issues of major import for our discipline, the reaction of American psychologists is quite frightening. What explains that reaction?

Americans are pragmatic. As long as papers are accepted for publication, why change.

From all your years of unflinching criticism of some misconceptions and practices rooted in psychology, are there moments that you recall with particular fondness or that you cringe when you think about them?

I have a special fondness for 4 findings that, at the time, were inconsistent with popular ideas. One fact was the discovery of the sudden increase in duration of fixation time at 7–8 months to stimuli that had been recruiting shorter attention spans from 3 to 7 months. I interpreted this as due to maturation of brain connectivity between temporal sites and the frontal lobes which enhanced the infant's working memory and the ability to relate the present event to a schema for a recent past event. This explanation of object permanence is inconsistent with Jean Piaget's interpretation.

A sweeter moment accompanied the data showing that the infant's cry to separation from the parent appeared and peaked at the same age in children from the USA, Israeli kibbutzim, a Mayan Indian village in Guatemala, the town of Antigua in Guatemala, and Kung San families in Botswana, even though the infants in these 5 settings experienced very different maternal practices in the first year. This evidence contradicted the popular idea that separation fear was a sensitive index of the infant's attachment to the mother. The data are on page 49 of *The Human Spark* (Kagan 2013).

A third moment occurred when my students and I discovered the emergence of four basic competences in the interval between 12 and 24 months. The four are the ability to infer the thoughts of another, a symbolic language, a moral sense and self-awareness. The close correspondence in time of appearance of all four is due partly to brain maturation in the second year (Kagan 1981).

A fourth pleasing discovery occurred in 1972 while I was observing infants and families in the village of San Marcos on Lake Atitlan in Northwest Guatemala. Although the one-year-olds were severely retarded and apathetic, because they received little stimulation as they lay in hammocks in the back of the adobe home, the behaviors and select cognitive skills of the five-year-olds, who were allowed to leave the home, resembled the traits of children of the same age in other societies. That observation in 1972 was inconsistent with the premise that an infant who did not receive adequate stimulation in the first year was at serious risk for later problems (Kagan and Klein 1973). This paper did attract some criticism, but later work by others affirmed our original inference.

Looking at what you do from the contrary positions, is there any particular type of criticism aimed at you that you feel is particularly serious and justified?

I have often made speculations with insufficient evidence. I also have been a bit biased in my reviews by emphasizing the papers that support my view and not giving equal time to the papers that do not.

Has this ever led to any problems? Have you ever had to distance yourself from theories or hypotheses you had formulated in a rush to judgment?

Yes. The concepts I used in "*Birth to Maturity*" (Kagan and Moss 1962) were too indifferent to age and gender.

As you wrote in your autobiography (Kagan 2007), **this book brought some degree of celebrity to you and to Howard Moss because you had discovered what the community wanted to believe. Is the rule still in psychology that these achievements are rewarded, which confirm what most of us believe in?**

Yes, that is true in all the sciences.

Do you still consider the discoveries written about by Judith Harris in *The Nurture Assumption* to be part of this group?

Yes.

Let's spend a moment discussing the problems of applied psychology, which you are also familiar with. The World Health Organization estimates that roughly 350 million people around the world suffer from depression, which is almost 5% of the population. On a global scale, over one-third of people at some time in their life meet all diagnostic criteria for at least one psychological disorder. These data suggest us that we are experiencing a pandemic of mental illness on a scale unknown in human history. The numbers illustrating the scale of mental disorders and illnesses are continually growing. What do you think about all this?

The high prevalence of a diagnosis of depression as a mental illness is due to the incorrect assumption that a bout of depression is abnormal and reflects a compromise in human brain functions. I suspect that depression was far more common during the Dark Ages when infections, premature deaths of infants and parents, bandits, fear of God's wrath, and hunger were common. Psychiatrists are not paying enough attention to the reason for a bout of depression. The medicalizing of anxiety and depression has led to the belief that we have an epidemic of depression or anxiety, when, in fact, there is an increase in only select causes of these states. For example, I suspect that social anxiety is more prevalent now than 1000 years earlier because so many live in large cities rather than villages. Moreover, unsureness over one's moral imperatives, which can precipitate a depression, is more common in young adults in 2018 than 1018. And as the number of adults older than 60 years increases, depression due to a chronic illness increases. The main point is a need to parse the term main depressive disorder (MDD) into the many different cascades that can trigger a depression and to stop treating the concept as a unity.

Just after you received your B.S. degree, the first edition of the *Diagnostic and Statistical Manual of Mental Disorders* (*DSM*) was published. It described a mere 106 illnesses and disorders. The most recent 5th edition from 2013 contains almost four hundred. In just 60 years, the number of mental disorders has multiplied nearly four-fold. If we are treating these numbers and their growth seriously, we are forced to conclude that humanity is facing total decimation and the world is going mad. Another conclusion that can be drawn from analysis of these numbers is that over a period of 60 years, psychopathology has made unprecedented progress by identifying the majority of psychological illnesses and disorders that have plagued civilization, and which had gone for so long unidentified. Or perhaps there is yet another explanation for this phenomenon?

The increase in DSM categories is due partly to the discovery of different patterns of symptoms for each DSM category. Second, psychiatrists have decided that behaviors or moods that were not a sign of illness 200 years ago are signs today. For example, in nineteenth-century America, men were supposed to be hyper-masculine and behave in ways that would be called borderline personality disorder today. The guilt over premarital sex in Colonial America was normal in the seventeenth century, not today. Children who were hyperactive and had a short attention span in medieval Europe would not have been diagnosed with an illness because these traits did not compromise adaptation in their rural setting. The point is that historical changes in circumstances alter the likelihood that a trait will or will not be called a sign of mental illness. The exceptions are severe intellectual retardation, incoherent thought and severe compromises in the ability to interact with others.

Does this mean that you consider psychological disorders and illnesses as products of culture?

No. Most DSM categories are heterogeneous combinations of biological vulnerabilities, experiential vulnerabilities from childhood, trigger events and the immediate circumstances when a trigger event happens to a person with a vulnerability.

Would it thus be accurate to state that the conditions we presently live in are conducive to the occurrence of these trigger events, and thus cause the more frequent presence of psychological disorders?

Yes.

Your views on some disorders, particularly ADHD, have been at the heart of some considerably significant controversies. Has your perspective changed to any serious degree, or is it still your view that ADHD is an invention?

ADHD stands for a variety of disorders with different causal cascades. Each cascade is real, but using the term ADHD for all of them is an invented idea.

What other conditions do you feel are invented and over-diagnosed?

Most DSM categories contain many different causal cascades.

An article published just this year (Meichenbaum, Lilienfeld 2018**) informs us that the number of psychotherapeutic modalities has already exceeded 600. Is their intensive development a solution to the problems we are discussing?**

The number of psychotherapeutic regimens has increased because there is no theory of remission or cure. A patient usually gets better if the healer and patient share the same ideas about cause and cure, as Jerome Frank wrote many years ago in "*Persuasion and Healing*" (Frank 1961). No therapy is better than any other for all patients with a particular symptom. This judgment does not apply to the small proportion of cases of autism or bipolar disorder.

Does that mean that psychotherapy is merely a ritual, and the various modalities it offers for explaining the causes of disorders and manners of treating them have nothing to do with what science has discovered so far?

Psychotherapy is an interaction between a patient and a clinician. Therapy is successful when patient and clinician share the same ideas of cause and cure, and the patient believes clinician is wise and cares about patient. The actual ritual is not very important when these two conditions hold. That is why shamans can cure.

Your definition is also suitable for recovered memory therapy. I've encountered many patients who have undergone such therapy, and they unanimously shared the views of therapists on issues of cause and cure, and they believed the clinician is wise and cares about their patient. I would not, however, consider the results of these psychotherapy interventions a success, and in many cases, they were catastrophic. Do you think that your understanding of psychotherapy could help in sanctioning similar modalities?

The failure of the therapy could be due to the severity of the patient's symptoms, the therapist's personality or the patient's evaluation of the therapist.

Can't this just be due to some incorrect theoretical assumptions made by psychotherapy?

Partly, yes, too many therapists believe their theory is best.

Probably we could spend a lot more time analyzing the problems inherent to our discipline. But let's set them aside for now, to take a closer look at psychology's achievements. Your studies and publications have reached the top of the mountain in terms of citations and are a fundamental part of our field's accomplishments. Which of your discoveries do you consider the most important in the development of science, and why?

Psychology has made many advances since its origin in the nineteenth century. The discoveries from my laboratories that generate pride are:

- The transition in the first year at 7–8 months, described earlier;
- The emergence of inference, language, moral sense and self-awareness in the second year;
- The fact that the retarded San Marcos one-year-olds recovered normal functions after being allowed to confront the world outside the home;

- The data from the Fels longitudinal sample implying that the prediction of many adult personality traits improves after age 6 years, summarized in "*Birth to Maturity*";
- The discovery of the two temperamental categories called high and low reactive.

The discovery of the three kinds of conceptual styles called analytic, relational and conceptual.

Some of these discoveries have led you to question the myth of infant determinism, itself firmly anchored not only in imagination of the general public but also in science and therapy, as you presented in your book *Three seductive ideas*. Do you think your arguments have convinced therapists, including child therapists, to abandon practices based on such assumptions?

No, I do not think so.

What about scientists? I'm curious as to how you judge the impact of your discoveries on the perspectives of other scientists. Have they come to accept your findings?

Many now accept the claim that for most children, excepting the rare cases of extreme deprivation in Romanian orphanages, the events of the first 3 years are not deterministic when the later years are benevolent.

Are your convictions also shared by adherents of Bowlby's attachment theory?

I do not think so.

Let's get back to the achievements of our field. The power of science is found in the questions it asks and the answers it provides. What other great questions have psychology managed to answer so far?

Psychologists have illuminated many puzzles. I regard the following as having high significance:

- Establishing that associations are formed when two events occur contiguously in time or space, or have links to related words or schemata;
- The growth of working memory in development;
- The special features of procedural, declarative and episodic memory;

- The role of the prefrontal cortex in modulating actions and feelings;
- The Gestalt laws of perception;
- The bases of color vision;
- The different outcomes that accompany variation in the social class of the family of rearing;
- The idea of "biological preparedness" for an event;
- The return of temperament as a contributor to personality;
- The growth of language in the first 5 years in different societies;
- The influence of culture on behavior, values and emotions.

Which great questions remain unanswered?
Many questions remain unanswered. For example:

- Why does class of rearing exert so much power?
- How do identifications with family, class, ethnicity and/or religion form?
- How does any psychological product emerge from a brain profile?
- How do genes, brain patterns and life history blend to produce a specific outcome in a person of a particular age, gender, class, ethnic group and local setting?

Do you think searching for answers to these questions can be aided by the intensive growth of AI that we are witnesses to?
I am skeptical that AI will be able to simulate human thought because AI relies only on a digital code while humans rely on blends of schemata, semantic forms and prototypes. Second, all AI programs require someone to assign a weight to each of the features in the events to be presented. These weights are often based on intuition and ignore the setting in which the events occur. But the human brain-mind alters the weights depending on the context.

Our conversation will also probably be read by young people just getting their careers started, and who are wondering what specialization to explore. What areas of investigation in psychology do you think are particularly promising and would recommend?
In my opinion, the areas ripe for study are: relation between brain states and perception, the effects of the context on behavior and brain

profiles, the relations between the experiences in the homes of advantaged and less advantaged infants and children in at least 3 different cultures and the child's cognitive skills and behaviors.

Who would you point them to as an example to follow?

Some good role models are David Hubel and Torsten Wiesel, Charles Darwin, Neal Miller, Endel Tulving, Alexander Thomas and Stella Chess, John Garcia, Russell Poldrack, Gyorgy Buzsaki, Jennifer Doudna, Vera Rubin, Eleanor Gibson, Nora Newcombe and Barbara McClintock. These scientists were open to new ideas, patient, careful, attended to details and brought passion to their work.

What should they be careful of? Can you give them any other advice?

Be careful of favorite ideas you want to believe are true, examine raw data before you compute any statistic, avoid covariance if possible, do not begin a study based on a word, instead study a puzzling phenomenon and do not rely on one predictor and one outcome.

Someone with your achievements and experience can allow himself to engage in a bit of futurology. In your opinion, what will psychology look like in two or three decades? Do you think that it will remain a field unable to overcome the crisis and escape the limitations you wrote about in *Psychology's Ghosts*, or will it manage to reach a higher plane?

I cannot predict what psychology will be like in 2048 because I do not know what new machines or procedures will be invented. The discovery of the structure of DNA because of X-ray crystallography changed genetics in a major way.

Do you think we will finally achieve a wide-ranging theory that incorporates the achievements of psychology into a uniform system of knowledge?

I do not believe this will happen in next 20–25 years because we need many more firm facts.

What are the greatest challenges facing psychology in the twenty-first century?

The greatest challenges facing psychology are: recruiting young investigators who have the traits of the those mentioned earlier; developing new methods; a willingness to study important phenomena that are difficult to measure, such as ethnic and class identity, a feeling of confidence that a major discovery is likely if one is patient, appreciates that asking the right question is more important than trying to have a publishable paper, and a willingness to reject a favored idea if the evidence demands it.

Selected Readings

Kagan, J. (1981). *The second year*. Cambridge, MA: Harvard University Press.

Kagan, J. (1998). *Three seductive ideas*. Cambridge, MA: Harvard University Press.

Kagan, J. (2002). *Surprise, uncertainty, and mental structures*. Cambridge, MA: Harvard University Press.

Kagan, J. (2006). *An argument for mind*. New Haven, CT: Yale University Press.

Kagan, J. (2007). *What is emotion? History, measures, and meanings*. New Haven, CT: Yale University Press.

Kagan, J. (2009). *The three cultures: Natural sciences, social sciences and the humanities in the 21st century*. New York, NY: Cambridge University Press.

Kagan, J., & Herschkowitz, N. (2005). *A young mind in a growing brain*. Mahwah, NJ: Erlbaum.

Kagan, J., & Snidman, N. (2004). *The long shadow of temperament*. Cambridge, MA: Harvard University Press.

Kagan, J., Snidman, N., Arcus, D., & Reznick, J. S. (1994). *Galen's prophecy: Temperament in human nature*. New York: Basic Books.

References

Frank, J. (1961). *Persuasion and healing: A comparative study of psychotherapy*. London: Oxford University Press.

Haggbloom, S. J., Warnick, R., Warnick, J. E., Jones, V. K., Yarbrough, G. L., Russell, T. M., & Monte, E. (2002). The 100 most eminent psychologists of the 20th century. *Review of General Psychology, 6*(2), 139–152.

Harris, J. R. (1998). *The nurture assumption*. London: Bloomsbury.

Kagan, J. (1981). *The second year.* Cambridge, MA: Harvard University Press.
Kagan, J. (2007). Jerome Kagan. In G. Lindzey & M. W. Runyan (Eds.), *A history of psychology in autobiography* (Vol. 9, pp. 115–149). Washington, DC: American Psychological Association.
Kagan, J. (2012a). *Psychology's ghosts: The crisis in the profession and the way back.* New Haven, CT: Yale University Press.
Kagan, J. (2012b, August 2). *What about tutoring instead of pills?* Der Spiegel. Retrieved from http://www.spiegel.de/international/world/child-psychologist-jerome-kagan-on-overprescibing-drugs-to-children-a-847500.html.
Kagan, J. (2013). *The human spark: The science of human development.* New York, NY: Basic Books.
Kagan, J., & Klein, R. E. (1973). Cross-cultural perspectives on early development. *American Psychologist, 28,* 947–961.
Kagan, J., & Moss, H. A. (1962). *Birth to maturity.* New York: Wiley.
Meichenbaum, D., & Lilienfeld, S. O. (2018). How to spot hype in the field of psychotherapy: A 19-item checklist. *Professional Psychology: Research and Practice, 49*(1), 22–30.
Shea, C. (2004, August 29). The temperamentalist. *The Boston Globe*. Retrieved from http://archive.boston.com/news/globe/ideas/articles/2004/08/29/the_temperamentalist?pg=full.

4

Michael I. Posner: Cognitive Neuroscience, Attention Networks and Training the Brain

Don't feel you must solve the problem just hope to make a contribution to the solution.

Michael I. Posner

T. Witkowski, *Shaping Psychology*,
https://doi.org/10.1007/978-3-030-50003-0_4

Once upon a time, there was a young man devoted to the study of what he believed to be the most romantic branch of science—physics. This young man came to the conclusion that he was not talented enough for such a subject and decided to enter the field of psychology. That moment of self-doubt was a stroke of luck for psychology, as it gained a devotee who could help smash through the limits of its knowledge. During his career, he has been called an experimental psychologist, a cognitive psychologist, a developmental psychologist, a neuroscientist, and probably some less flattering things as well. With tremendous nonchalance Michael Posner has traversed the borders of particular specializations, his mind occupied only by seeking an answer to the question of how our brain functions. And the pieces of that puzzle could not be found in one narrow subdiscipline.

A truly great mind is needed to perceive the paths leading to answers in places where many would observe merely a curiosity to be quickly forgotten about and most would see nothing. The clue picked up by Posner that led him to breakthrough discoveries in psychology was the result of a certain study demonstrating that people needed more time to correctly match two letters which had the same name and the pair were upper- and lowercase (e.g., Aa) than when they were matching "AA" or "aa". The 80-millisecond increase that occurred for matching "Aa," which wouldn't register with most of us, led Posner to ask questions about the essence of how our brain functions; the answers he found have placed him in the pantheon of the greatest living psychologists. But let's start from the beginning...

In 1963, he was a young assistant professor at the University of Wisconsin in Madison. The department was run at the time by Harry Harlow, David A. Grant and Wilf Brogden. They were important in psychology and thought that anyone in their department should be as well. Posner planned to study attention. And now we return to the story about "aA" and "AA." In the course of his research, he observed something quite different than he had expected, and, from his perspective, something of greater importance. Indeed, it does take longer to match an auditory digit following a visual one than to match two identical auditory or two identical visual digits, just as he had predicted. However, when the task was changed from matching identical digits to one of determining whether digits were odd or even, he observed that time was no longer needed to switch between modalities. Instead, he found that whenever the two items were not physically identical ("Aa" vs. "AA"), the time increased. From his perspective steeped in the learning theory of the 1960s, this was a genuine puzzle. To him, this indicated that some internal processes are involved in determining the difference.

It was then he recalled that about 100 years ago the great Dutch physiologist Franciscus Donders had first measured the time needed for the covert mental processes of recognition and choice by using the subtractive method. Although he was aware of criticism of the subtractive method by a number of psychologists, he didn't hesitate to use it in explaining the brain's functioning. His paper describing these findings appeared in the *Psychological Review* in 1967.

Ten years later, in *Chronometric Explorations of Mind*, Posner (1978) applied the subtractive method devised 110 years earlier by Franciscus Donders to the study of several cognitive functions such as attention and memory. The subtractive method is based on the assumption that mental operations can be measured by decomposing complex cognitive tasks into sequences of simpler tasks. The method assumes that the effect of each mental operation is additive, and that it is possible to isolate the effect of a single mental operation by comparing two tasks that differ only by the presence or absence of that mental operation. Posner applied the same subtractive principle to the study of attentional networks using Positron Emission Tomography (PET), a neuroimaging technique that produces three-dimensional functional maps of the brain. In *Images of Mind*, co-authored with Marcus Raichle in 1994, Posner investigated the brain localization of cognitive functions by examining patterns of brain activation during progressively more complex cognitive tasks.

Chronometric methods opened up new opportunities for psychology when they were applied in the study of patients with neurological damage. This approach fits perfectly with the longstanding ambition of psychology to account for the relationship between mind and brain. It provided a foundation for the establishment of cognitive neuroscience, a field in which Michael Posner is recognized as a founding figure. Since that time, psychologists, neuropsychologists and cognitive scientists in general have profoundly changed their approach to research.

Neuroimaging has offered a series of increasingly sophisticated methods with which to explore the neural basis of cognition.

At the same time as he was making these discoveries, Posner began his exploration of higher-level issues. He started examining developmental and social issues and initiated a series of studies of temperament, testing the assumption that personality factors impact perceptual and cognitive processes. In particular, he investigated the development of executive attention during the early years of life. Later on, he became interested in genetics and took steps to bring together researchers who would contribute to the new field of cognitive genomics, investigating molecular genetics of self-regulation and its relationship to environmental influence. Exploring aspects of interest to a wide range of experimental and clinical psychologists, he showed that cultural factors (such as the quality of parenting) interact with individual temperamental factors (such as impulsivity and sensation-seeking). By doing so, he achieved a novel combination of genetics and intervention studies, which opened stimulating and socially relevant new fields of research. This is well illustrated in the book co-authored by Michael Posner and Mary Rothbart (2007), *Educating the Human Brain*, a major contribution to our understanding of brain development and its relation to education.

Professor Posner, you started your career at a time when psychology was growing beyond the previously dominant behaviorism. Essentially, you were one of those responsible for initiating this trend. What did the departure from behaviorism look like; did it ride a wave of rebellion and discord, or did it rather take on an evolutionary form?

I was not a part of the rebellion, if there was one. I have always felt it most important to build psychology in a cumulative way. Not to view either behaviorism or cognition as a full story, but as a part of how we expand psychology to account for what people both think and do. This was perhaps not a majority view, but I had come to psychology from a physics major and from working in a human factors group, from this perspective I hoped to contribute a part to what might someday be a full science of the functions of the human brain. I do think my views are supported because cognitive science does incorporate many of the findings of behaviorism for example in connectionism.

Your research contributed to the emergence of cognitive psychology. Its blossoming brought great hopes for the emergence of a wide-ranging theory, so lacking in psychology, that would facilitate the unification of research results across various areas of psychology and systematize our diffuse knowledge. Meanwhile, despite the incredible achievements of this field, wouldn't you say we are still quite far from the appearance of such a theory? I have the impression that we psychologists have even stopped talking about it in recent times.

I have taught psychology for more than fifty years. At the start it was a somewhat dull discipline built on the narrow direction of how rats turned in mazes but pretending this could be generalized into a full theory of humans. Today we remain far from a general predictive-explanatory theory of the human brain and behavior, but it is genuinely exciting to teach psychology. From how most people make economic decisions, to pictures of the brain as it tries to control anger, from how genes contribute to control of behavior to how learning shapes gene expression, the field is alive with ideas and results that interest students and others. Whether we will ever reach a full theory to help us understand human behavior we are already shaping the algorithms by which robots learn to approximate our behavior. To me it seems an amazing journey and I envy your opportunity to tell it.

Indeed, almost on a daily basis our field is generating new and interesting results that the media are eager to write about and discuss. However, is it perhaps the case that the absence of a necessity to maintain a theoretical consistency with a more general predictive-explanatory model of behavior leads researchers to compete to publish the most "sexy results", which themselves frequently prove impossible to replicate, and the career of a scientist is fashioned along similar lines to that of a media celebrity?

To some extent this is encouraged by journals that emphasize interest of the broad community, instead of contribution to the basic science of psychology. Interesting articles are important but contribution which build the underlying field are equally important. It is difficult to find the right mix of the two.

The results of your recent research have brought fascinating discoveries related to the possibilities of improving our mind, which are exciting quite a number of people. What are the clear conclusions that can be drawn from that research today?

We can certainly improve our minds by learning new facts and skills in diverse fields. When we have a strong semantic memory for a domain we can learn more easily and faster. There is little dispute about that, but can we change our brain in a way which improves the capability of the mind to learn? This is the unsettled question on which I and many others are working. My approach is to try to understand the mechanisms by which tasks improve with practice in hopes, only partly realized, that this would illuminate what it means to change the brain and how best can this be accomplished.

In an interview several years ago you also announced a longitudinal study on the effectiveness of attention training in children. Have you obtained any results?

We have obtained many important results in this research. We found that attention training at age 4–6 can improve the brain network involving the anterior cingulate cortex (ACC) as measured by EEG. The training results in improved IQ performance and improved ability to tolerate delayed rewards. We also found evidence that as early as 7 month there is evidence that the ACC portion of this network is active in error detection, however, we showed behaviorally that although errors are detected very early they are not acted upon in terms of slowing on the next trial until three years later. Resting state fMRI studied by others have also found evidence of early development during infancy of the ACC-mid prefrontal node but it is not connected to motor systems until much later.

In your public statements, you frequently emphasize the fact that we humans can regulate our thoughts, emotions, and actions to a greater degree than other primates. The results of your research show that perfecting one of our brain functions, such as attention, has a broad impact on self-regulation processes. I think that knowledge about this subject is not at all widespread. Could you perhaps expand a bit on your view of self-regulation?

Self-regulation is the ability to use our intention and goals to resolve conflict between competing responses. It appears that the executive attention systems serves most of this role in adults. Research of many people has identified a network including the anterior cingulate/mid prefrontal cortex, anterior insula and striatum which seems to resolve conflict to implement coherent behavior. This network is partly available at least by 7 months but develops mainly in early childhood up to about 7 years. It serves as a core network for self-regulation and individual differences in its development predict important aspects of adult outcomes.

The majority of readers are in all certainty asking themselves: can we—and to what extent—impact the formation of this network among children?

I believe we must try to do the research to accomplish this difficult goal. That is much of what I am trying to do these days. It has led to mouse studies to examine the basis for brain change and human studies to see if we can erect method to help development based on this basic understanding of development.

Your studies and publications occupy the highest positions in the citation indexes, and the results of your research are the foundation of most psychology textbooks. Which of your discoveries do you consider the most important in the development of science, and why?

I would not say these are really my discoveries but I played a role in the development of brain imaging and in the understanding of brain networks of attention.

Is there any particular type of criticism aimed at your work that you feel is particularly serious and justified?

Our imaging work was often criticized for its over emphasis on localization of function. The network approach has reduced that over emphasis by stressing the orchestration of multiple brain areas within and across networks.

Alongside the respectable research done, for example, by your team, brain improvement is also part of the extremely fashionable and commercialized fad of self-improvement, in which we can also find a great deal of pseudoscience and myths. On several occasions I have observed how the authors of heavily simplified marketing messages supported their claims with your research results. I think

it is worth trying to separate what is proven, real knowledge from bankrupt promises. Which of the most common claims regarding the improvement of our brains do you consider particularly harmful, and why?

I covered one of these, attention training in children in question 4. However, we also used a form of mindfulness meditation, which improved attention and after two to four weeks seemed to change white matter surrounding the ACC one of the key nodes of attention. We developed a mouse model to test the idea that the way meditation changed white matter was due to an increased frontal theta. Mice showed improved white matter after a month of theta stimulation. We also found reduced anxiety in the mice. I am now working to see if we can produce these effects by use of scalp electrodes and if so what behavioral changes will be produced.

These studies are indeed fascinating, and I have observed their use in marketing what is presently the very popular mindfulness meditation, but I don't quite understand why you consider this particularly harmful?

I do not consider mindfulness or its advertising harmful, but I do think we need to understand the mechanisms involved. Let us accept that meditation can lead to white matter change, however, that occurs only in the pathways related to executive attention. If we understand that this change is due, for example, to theta stimulation we can improve white matter anywhere in the brain. That is one reason why further experimentation of the mechanisms of meditation is important.

Your team also conducts research on epigenetics. There are, similarly to the field of brain improvement, many misunderstandings here. With the help of epigenetics, outworn New Age concepts are being revived, some therapeutic practices are being justified, and attempts are even being made to revive Lamarckian ideas. Various contemporary gurus commonly employ the concept of transgenerational epigenetic inheritance to sell us on practices that involve the therapy of inherited traumas, or to work on changing our own genome. How much truth is there in any of this?

Our main work has involved candidate genes in relation to the network underlying attention. As a part of this work we found that alleles of the MTHFR, gene, which can carry out the process of methylation with varying efficiency, are related to the speed of learning and executing various cognitive tasks. One interpretation of our findings with alleles of the MTHFR gene is that our results are due to an epigenetic effect. However, other interpretations are possible. While epigenetic effects certainly occur and influence the expression of genes in a given individual, the extent of their influence in various aspects of human behavior are still pretty unclear. Moreover, as far as I know there is no evidence they modify the genes in a way which is transmitted to the next generation, which might support Larmarckian views.

And the claims that we can modify our genotype and that of our future offspring through diet, meditation, etc., are also false?

Yes as far as I know in all these cases only the ability to influence genetic expression has been establish not modification of the inherited genome.

You have been a leader in the development and application of modern neuroscience techniques. Cognitive neuroscience is presently the subject of intense media attention, while at the same time of harsh criticism. I'd like to hear your thoughts on some of those critical voices. Let's begin with the famous dead salmon, which was examined by Craig Bennet and his team (2010), with the results presented in "*Neural Correlates of Interspecies Prospective Taking in the Postmortem Atlantic Salmon: An Argument for Multiple Comparisons Correction.*" Do you agree with him that the discoveries of contemporary neuroscience are frequently the same kinds of artifacts that his team identified in the brain of a dead salmon?

The work you cite is certainly correct in raising caution about multiple comparisons. As in all research it is important to have clear hypotheses which can be tested. The use of clear hypotheses which can be supported or not and replication of findings between separate experiments are the main ways of avoiding errors due to multiple comparisons. This is certainly true of all research but brain imaging because it can involves many brain areas is likely more susceptible to these errors that other areas of research.

This criticism is linked to another loud voice, also addresses to social neuroscience research. I am referring to the allegations made

by Edward Vul and his team (2009) **in an article originally published under the title "Voodoo Correlations in Social Neuroscience." What do you think of the issue of nonindependent analysis in behavioral neuroimaging? Are most correlations really artificially inflated as a result of spurious statistical procedures?**

I read the Vul article and the reply by Lieberman, I do believe that the high correlations reported in many neuroscience articles are misleading. This is partly because people have little intuition of how variable correlations with small n can be. So, a value of .7 may be as low as .4 or as high as .9. MRI also has the problem of multiple comparison which need to be considered by the reader.

In 2016, there was a tremendous outcry in connection with the detection of an error in fMRI analysis software. In the media there were suggestions that the results of research from the previous 15 years may prove worthless (Crew 2016). **In your view, was this panic justified, and how was the problem solved?**

The algorithms used in MRI studies are complex and of course errors arise in their construction and use. However, convergence between cellular recording, behavior and MRI shows that everything learned in the last 15 years is certainly not wrong.

I know that putting the question in this manner is asking for a significant simplification, but it is one frequently posed by readers confronted with the type of information presented above: What percent of the entire collection of results of neuroscience studies can we safely assume are correct and verified?

About 0% are verified in the sense they could not be changed by further studies, but probably more than 80% can be replicated under the conditions and with the organisms that were originally studied.

A somewhat different kind of criticism of the manner in which research in cognitive neuroscience is conducted is that of William Uttal (2001) ***The New Phrenology*. He questions the very idea of location and dismisses its rebirth in modern neuroscience. How do you perceive his arguments?**

I had the pleasure of debating this issue with Bill Utall several years ago at Northwestern University. I personally think that the brain network approach has reduced the emphasis on a single brain area as the source of effects found in behavioral experiments. Of course as Bill pointed out localization is not everything, but clear progress has been made with the network approach and its implementation in deep learning algorithms has provided a connection to important applications.

I am glad that you mentioned deep learning algorithms. We are fortunate to be witnesses to the remarkable development of AI. Does it help us in finding answers to important scientific questions?

Certainly, having a method to bring together new findings in algorithms that can be applied to robots etc. greatly enlarges our field and provides an important method for attracting brilliant new minds to it. These are crucial to the development of any field. I believe AI has provided an extremely important application of what is being found out in psychology. The work of Rumelhart, McClelland and Hinton has been basic to current uses of AI but the role of psychology in these developments have not been well recognized (Rumelhart et al. 1986).

What was the cause of this?

Generally everyone loves a new name, like cognitive neuroscientist, computer scientist or AI researchers. There is no harm in that but when people say there is no progress in psychology part of the reason is that when there is, it is given a new name.

In your opinion, what important questions has psychology already answered?

It may not be useful to put the question in this way. But there is clear progress. Consider that 20 years ago it was generally thought that infants were unable to process information, now we know that the new born infant has the capability of distinguishing all of the worlds phonemes, and have a clear theory (Kuhl et al. 2007) on how the infant auditory system at 7–12 month absorbs experience and how exposure to the native language shapes the phonemic system. Of course in science there are no final answers, but among the questions currently pressing are (1) how well neural networks can capture the full range of human functions, (2) how can brain networks best be related to cellular and molecular levels of analysis, (3) what are the limits of conscious control (4) can we improve brain mechanisms

Which methodological problems in our field need to be solved as quickly as possible?

A method to control cells in the human nervous system non-invasively would carry us a long way forward.

I think that a person who has participated in the most critical breakthroughs in the history of our field can afford to share his vision of the future with others. How do you imagine our field of science over the next two or three decades?

I think the connection between the network, cellular and molecular levels of analysis is most critical for the field. If we can solve it we will have a deeper understanding of brain function and a better chance to help people with mental disorders.

Many young people at the threshold of their careers are looking for advice on what direction to take with it. What would you tell them?

Everyone should to choose a large central problem about which they are excited to solve and for which they feel they could make a contribution. Even if you should be wrong you at least have the satisfaction of knowing you tried. Don't feel you must solve the problem just hope to make a contribution to the solution.

That's a quite romantic-sounding piece of advice. I recall one of your statements in which you complained that while applying for research grants, usually you had to hide your intention in order to get funding support. Has anything changed since those times? Perhaps we could also give our readers some practical advice for how they can secure funding to seek solutions to problems that are of interest to them?

I don't think that the limited vision of many funding organization has changed, but my best advice is as follows: In securing funding you need to be sure you give the funding agency a clear vision of where you are headed as well as specific information on your current proposal. You will need colleagues at your institution who can review your request and help you understand how to convey the information to others. You must be prepared to revise your proposal based on the advice you get from colleagues and from the funders.

Who would you name as a model to follow for young people just getting their careers off the ground?

There is really no model for an individual's career. We all differ in goals, strengths, weaknesses and interests. Fortunately in the diverse US educational world and even more diverse economy most people are successful. Keep in mind you never know for sure there is no appropriate control group.

Can you give them any other advice?

You should join at least one group of like-minded psychologists interested in aspects of the field similar to your interests. These may include: The Association for Psychological Science or the American Psychological Association also consider Cognitive Neuroscience Society and other specialty groups. I have written many articles and book chapters providing details of many of the things I have discussed in this interview. I keep a complete CV and some more recent papers at my blog. I have posted in December 2018 and should update soon. https://blogs.uoregon.edu/mposner/.

A final question. While working on this book, I asked my readers to submit one question they would like to ask the most eminent living psychologists. I received 30 of them. Would you agree to answer some of them? If so, please choose one or two numbers from 1 to 30.

I am happy to try to answer question 17 and 26.

So, question number 17 is: What questions are you asking yourself?

I am working on methods to change the pathways that connect the attention networks to memory, emotion and thoughts. I hope but am by no means sure this may be of aid many people.

And the question number 26: The idea in psychology which made the most good?

The idea that the human brain is changed in crucial ways by experience and we can even image the changes, provides new impetus to improving the human condition. To me this is a very crucial development in the field.

Selected Readings

Mayr, U., Awh, E., & Keele, S. W., (2005). *Developing individuality in the human brain: A tribute to Michael I. Posner*. Washington: APA Books.

Posner, M. I., & Boies, S. J. (1972). Components of attention. *Psychological Review, 78*, 391–408.

Posner, M. I., & Keele, S. W. (1968). On the genesis of abstract ideas. *Journal of Experimental Psychology, 77*, 353–363.

Posner, M. I., & Petersen, S. E. (1990). The attention system of the human brain. *Annual Review of Neuroscience, 13*, 25–42.

Posner, M. I., & Rothbart, M. K. (1998). Attention, self regulation and consciousness. *Philosophical Transactions of the Royal Society of London B, 353*, 1915–1927.

Posner, M. I., & Rothbart, M. K. (2000). Developing mechanisms of self-regulation. *Development and Psychopathology, 12*, 427–441.

Posner, M. I. (1989). *The foundations of cognitive science*. Cambridge, MA: MIT Press.

Posner, M. I., (2012). *Attention in a social world*. Oxford: Oxford University Press.

Posner, M. I. (Ed.). (2004). *The cognitive neuroscience of attention*. New York: Guilford.

References

Bennett, C. M., Baird, A. A., Miller, M. B., & Wolford, G. L. (2010). Neural correlates of interspecies prospective taking in the postmortem Atlantic Salmon: An argument for multiple comparisons correction. *Journal of Serendipitous and Unexpected Results, 1*(1), 1–5.

Crew, B. (2016, July 6). A bug in FMRI software could invalidate 15 years of brain research. *Science Alert*. Retrieved from https://www.sciencealert.com/a-bug-in-fmri-software-could-invalidate-decades-of-brain-research-scientists-discover.

Kuhl, P. K., Conboy, B. T., Coffey-Corina, S., Padden, D., Rivera-Gaxiola, M., & Nelson, T. (2007). Phonetic learning as a pathway to language: New data and native language magnet theory expanded (NLM-e). *Philosophical Transactions of the Royal Society of London. Series B, Biological sciences, 363*(1493), 979–1000.

Posner, M. I. (1978). *Chronometric explorations of mind*. Hillsdale, NJ: Lawrence Erlbaum Associates.
Posner, M. I., & Mitchell, R. F. (1967). Chronometric analysis of classification. *Psychological Review, 74*, 392–409.
Posner, M. I., & Raichle, M. E. (1994). *Images of mind*. New York: Scientific American Books.
Posner, M. I., & Rothbart, M. K. (2007). *Educating the human brain*. Washington, DC: APA Books.
Rumelhart, D. E., McClelland, J.L. & the PDP Research Group (eds.). (1986). *Parallel distributed processing: Explorations in the microstructure of cognition, Vol. 1: Foundations*. Cambridge, MA: MIT Press.
Uttal, W. R. (2001). *The new phrenology: The limits of localizing cognitive processes in the brain*. Cambridge, MA: MIT Press.
Vul, E., Harris, C., Winkielman, P., & Pashler, H. (2009). Puzzlingly high correlations in fMRI studies of emotion, personality, and social cognition. *Perspectives on Psychological Science, 4*(3), 274–290.

5

Scott O. Lilienfeld: Clinical Psychology, Evidence-Based Treatments and Skepticism

We have a long way to go, but at least we are on the road to salvation.

Scott O. Lilienfeld

T. Witkowski, *Shaping Psychology*,
https://doi.org/10.1007/978-3-030-50003-0_5

When I was a little boy, most of my friends, myself included, went through a period of wanting to become firefighters, but none of us thought about becoming fire safety inspectors. Our dreams, I suppose, were the product of our environment. Probably there never was a fire safety inspector on the front page of the newspaper, even though it was because of the hard work he did in his region that not a single fire had erupted in 25 years. Fame and fortune were and still are the domain of the firefighter who rescued a helpless child from a burning home. It is rare, if at all, that hard-working and cautious people become heroes, in particular those who clean up other people's messes—but it is often because of these people that any progress is made at all. One such potential hero in the field of psychology is Scott O. Lilienfeld.

In both psychological science and applied psychology, there is a big mess. Over 600 therapeutic modalities and literally thousands of theoretical constructs need people to come in and tidy up. Lilienfeld is in the front ranks of scientists prepared to take on the challenge of tidying up psychology, particularly clinical psychology. It is no exaggeration to say that he is the world's most well-known skeptic psychologist engaged in demasking pseudoscience in our field.

Considering his broad range of interests, Scott Lilienfeld describes himself as a generalist. This is a very rare attitude in an age of hyperspecialization, but perhaps it is precisely for this reason that he could give himself over to education and the popularization of science. Fascinated by Carl Sagan, whom he had the fortune to meet in person and listen to his lectures, he spends a great deal of his energy on learning critical thinking and a scientific understanding of reality. The fruits of this work include several highly popular books. In *50 Great Myths of Popular Psychology: Shattering Widespread Misconceptions About Human Behavior*, written with Steven Jay Lynn, John Ruscio, and Barry Beyerstein, Lilienfeld (2010) examines 50 common myths about psychology and provides readers with a "myth busting kit" to help learn critical thinking skills and understand sources of psychological myths, such as word of mouth, inferring causation from correlation, and misleading film and media portrayals. Translated into multiple languages, it serves as an educational guide to critical thinking about psychology.

No less famous is his work *Brainwashed: The Seductive Appeal of Mindless Neuroscience* written along with his colleague Sally Satel (2013). It focuses in its entirety on demasking the misleading falsehoods and simplifications of neuroscience. The authors target such practices as functional magnetic resonance imaging (or neuroimaging) to "detect" moral and spiritual centers of the brain, which they call "oversimplified neurononsense."

Lilienfeld is an exceptionally prolific author, and his publications inevitably generate much controversy and protests among scientists and practitioners devoted to the very things he criticizes. He has taken pieces out of believers in the Rorschach Inkblot Test, he has written critically about eye movement desensitization and reprocessing (EMDR), recovered memory therapy and misconceptions in autism research, such as the MMR vaccine controversy, as well as fad treatments such as facilitated communication.

Lilienfeld founded the *Scientific Review of Mental Health Practice*, the only peer-reviewed journal devoted exclusively to distinguishing scientifically supported claims from scientifically unsupported claims in clinical psychology, psychiatry, social work and allied disciplines. For several years, he also served as its editor-in-chief. Unfortunately, the journal ceased operations for the simple reason of insufficient funding.

He is known to take a clear and unequivocal position in controversial public debates concerning psychology. One example was during the most recent presidential election, in which Donald J. Trump campaigned, and there was much discussion of abandoning the Goldwater Rule; since the 1970s, this rule has forbidden psychiatrists and psychologists from issuing public diagnoses of politicians and other famous people. Lilienfeld came out in favor of abandoning this rule, while at the same time admitting that in most cases such diagnoses are of far less value than direct observation of concrete behaviors engaged in by people in the public eye.

Professor Lilienfeld, you're one of the world's most well-known psychologists engaged in unmasking pseudoscience in the field of psychology. What led you go beyond your clinical research and engage to such a significant degree in countering myths and misconceptions in our field?

I love and value the field of clinical psychology, and when I entered graduate school in the early 1980s, I assumed—naively—that most practitioners in this field were relying heavily on science to inform their clinical practice. It was not until a decade or so later that I belatedly became convinced that this was not in fact the case, and that much of our field was underpinned by questionable science. I was surprised by the extent to which many clinicians were relying on their intuitions and clinical experience rather than on data. My graduate education at the University of Minnesota, wonderful as it was in many ways, had not adequately prepared me for this sobering realization. It was in the mid-1990s or so that I then decided to make the challenging of myths and misconceptions in psychology a major focus of my work.

Unmasking others probably doesn't make you a lot of friends, does it? How are you and your work in exposing pseudoscience perceived by other psychologists?

I don't know, but I suspect that the reaction is bimodal. I've certainly made my fair share of enemies, as you note, but I'm also made quite a few close friends along the way. I don't mind tough criticism just so long as it remains substantive and doesn't become personal. Being controversial comes with the territory.

What kind of criticism directed at you do you think is justified?

I try my best to take all substantive criticism seriously. I'm particularly open to the possibility that I have been prematurely dismissive of some previously unsubstantiated claims, and welcome corrective evidence to the contrary. For example, in the early 1990s, I raised questions regarding the efficacy of eye movement desensitization and reprocessing (EMDR) for posttraumatic stress disorder and allied conditions, and I've since become persuaded that EMDR is indeed somewhat efficacious for these conditions. Still, I remain unpersuaded that it is more efficacious than standard cognitive behavioral therapy, although some recent data raise this possibility.

In listening to interviews with you, I also find myself wondering how you are able to maintain a healthy emotional distance from all of those abuses you encounter. Don't they infuriate you, don't they throw you off-balance? After all, so many of them cause real harm and human suffering.

I do occasionally become angry and enraged, and in those cases, I've learned to step back a bit, at least temporarily, as I worry that my emotions may impair my ability to evaluate assertions impartially. I've also learned over time to try to channel my negative emotions into constructive avenues, such as (hopefully) informed scientific criticism. Outrage can be a healthy emotion so long as you can control it. If it begins to control you, it can impede your approach to science.

Your attitude toward pseudoscience is untypical of the academic community. Many of them turn a blind eye to pseudoscience practices. Why, in your opinion, is this so?

There are surely several reasons. One is that our field doesn't greatly value efforts to challenge pseudoscience. By and large, the reinforcements tend to go to those who publish lots of papers and obtain big grants, not to those who question popular claims. A second is that one inevitably makes a number of enemies, as you've already noted. Many professors are conflict-avoidant and would understandably prefer to do their own thing, without having to incur the wrath of others. Third and finally, academia is a bit of a cocoon. Many professors don't appreciate the magnitude of pseudoscience, because they rarely encounter fervent believers or practitioners of pseudoscientific claims in their everyday interactions with colleagues and students. They need to spend more time out of the halls of the Ivory Tower.

A consequence of what we're discussing is that the myths of psychology that you unmask include not only those that have arisen by some strange set of circumstances, or that have become received wisdom as a result of distortions by the media, or purposefully created by people trying to make them into money-spinners. There are also many that have been established by scientists who failed to remain skeptical enough. Which of those do you think are the most dangerous, and why?

The myths that I perceive as most dangerous are those that have the potential to damage lives. For example, the misconception that memory operates much like a video camera or tape recorder can lead jurors and other triers of fact to accord undue weight to confidently expressed eyewitness testimony, thereby landing innocent people in prison. It can also lead people to accord uncritical acceptance to recovered memories of child sexual abuse. As another example, the erroneous belief that the polygraph test is a largely infallible detector of lies has ruined many careers and hurt countless innocent people. Another class of myths that I see as pernicious is those that lead us to falsely appraise the key influences on our emotional well-being. For example, many people massively overestimate the extent to which their happiness is attributable to life events as opposed to their interpretation of these events. Indeed, research suggests that our long-term happiness is often surprisingly independent of what occurs to us. As a consequence, many of us spend large chunks of our lives trying to change our life circumstances, such as trying to find a better job or better place to live. Doing so can sometimes be helpful if one's occupation or home is grossly suboptimal, but in many other cases it will be more helpful for us to find ways of viewing our present life circumstances in a more positive light.

Are we at all able to examine something like happiness and suggest people the ways to get to it? Does not it stand somewhat in contradiction with the Hume guillotine?

By Hume's guillotine, I assume that you mean the hoary "is-ought" problem. Despite what some prominent philosophers have suggested, I remain persuaded that one can't logically extract an ought from an is. Hence, with respect to happiness, I suspect that science may be able to shed light on how to achieve it, and it can certainly shed light on its correlates. I am much more dubious, however, that science can answer the question of when we *ought* to foster happiness. That's a value-laden judgment that lies outside the realm of science. In some situations, such as being in a concentration camp or victim of terror, it would typically be maladaptive to be happy. Again, science can sometimes inform large-scale societal decisions about how best to achieve happiness; but the question of whether and when to foster happiness invariably entails complex value judgments that are a-scientific.

Some of very popular myths are created and spread within highly regarded neuroscience. With Sally Satel you devoted the whole book to describe them. Has anything changed since the publication of this book?

Since our book was published, we've seen something of a backlash against overstated neuroscience claims, a trend that I basically see as healthy. I say "basically" because at times this backlash has itself become overstated, with some authors going so far as to decry most or all neuroimaging findings as akin to phrenology or other pseudosciences—an assertion I forcefully reject. At least among my colleagues in clinical psychology, I'm beginning to see a somewhat more thoughtful and scientifically balanced perspective on neuroscience claims, with a recognition that neuroscience can provide one extremely helpful lens of analysis on human behavior without displacing other lenses. I suspect that some of this recognition has come from the realization that many of the most expansive promises of neuroscience haven't been borne out. When I was in graduate school in the 1980s, many prominent leaders in the field were confidently predicting that neuroimaging would supplant—not merely supplement—psychiatric diagnostic tools. They assured us that within a matter of a decade or so, brain scans would replace psychiatric interviews and questionnaires as a means of diagnosing schizophrenia, depression and other serious mental disorders. Needless to say, that hasn't happened; when DSM-5 was published in 2013, it contained not a single neural marker among its many hundreds of diagnostic criteria. We've seen a similar pattern of hype, followed by disappointment and disillusionment, in a number of neuroscience domains over the past few decades. As a consequence, many psychologists have rightly become more skeptical.

Among skeptics you are also known as a person who writes and speaks about the need for better communication between scientists, skeptics and the general public. In your opinion, what is the key to the minds possessed by totally wrong pseudoscientific ideas?

The key is recognizing that it's not an "us versus them" proposition. We are all prone to fallacious thinking, myself included, and scientific thinking doesn't come naturally or easily to any of us. So it's critical to display respect to individuals who hold questionable beliefs, as we've all been there ourselves. I've sometimes been surprised, even taken aback, at the extent to some of my fellow skeptics exhibit blatant disdain or contempt toward individuals who sincerely hold dubious beliefs (I'm not talking here about outright charlatans who know full well that they disseminating nonsense, as I have no patience for them). Again, the key is to communicate respect and intellectual humility. We scientists typically aren't very good at this, and we are rarely trained in it either. I have no doubt that I have failed repeatedly in this regard, although I hope that I am getting a bit better with each passing year.

The clearly pseudoscientific practices we are discussing are one thing, and some scientific practices that are leading our field over a cliff into something that can no longer be called science are something totally different. Recently, the problems caused by such practices have been growing in intensity. What I have in mind is fraudulent research, methodological carelessness resulting in studies that are essentially non-replicable, the lack of access to raw data and other phenomena that are detailed in the book *Psychological Science Under Scrutiny*, which you edited. Taking all of this into consideration, would you feel it is fair to say that psychology is presently experiencing a crisis?

Yes, although I don't see the crisis as limited to psychology. I see it as a far broader challenge confronting science at large. We are witnessing similar problems of replicability in a number of other domains, such as economics, medicine (including psychiatry) and cancer treatment, among others. I suspect that psychology is in the crosshairs of public opinion largely because we've been the field that has taken the hardest look at itself. I actually see these as encouraging days for psychology, and I concur with those who contend that we undergoing a renaissance at least as much as a crisis. We are operating the way a healthy science operates; we are identifying errors in our standard ways of doing business and doing our best to root them out and correct them. We have a long way to go, but at least we are on the road to salvation.

I've invoked the subject of crisis on purpose, because I am under the impression—and not for the first time—that scientists confronted with improprieties in their research field are merely acting as though they are extremely surprised. After all, Jacob Cohen (1962) has been writing and speaking about the problems resulting from null hypothesis significance testing since the beginning of the 1960s. At that same time, Leroy Wolins (1962) initiated a discussion about access to raw data. Other problems have been under discussion for decades now, but action is only taken when the media take

an interest. Why is this the case when the majority of them result from insufficient scientific skepticism, which is the essence of doing science?

I suspect that reinforcement contingencies provide much of the answer. In the words of Upton Sinclair, "It is difficult to get a man to understand something when his salary depends upon his not understanding it." In the case of academic psychology, we are rewarded primarily for publishing lots of papers in high-profile journals and securing large federal grants. There are few or no incentives for us to engage in soul-searching regarding our cherished methodological assumptions. In fact, there are often active disincentives, because collecting larger samples, conducting multiple studies, avoiding p-hacking techniques, tempering one's claims, and the like, are often implicitly discouraged. (I recently received a review of a manuscript in which the reviewer criticized us for spending too much time acknowledging potential weaknesses in our theoretical position.) Investigators who make concerted efforts to engage in such safeguards often find it harder to get job offers, and to receive tenure and promotion after they are hired. Fortunately, this state of affairs is gradually changing in many quarters.

Which of the issues addressed in *Psychological Science Under Scrutiny* do you think are the most serious, and why do you think so?

Self-delusion, including confirmation bias. I suspect that most researchers are fundamentally honest. But I also suspect that many or most of them are largely unaware of the extent to which their theoretical allegiances and hypotheses can deceive them, leading them to engage in methodological practices, such as p-hacking and HARKING (hypothesizing after results are known), which can in turn lead them to spuriously corroborate their beliefs. Adding to this problem is what Emily Pronin and colleagues (2002) term "bias blind spot," which afflicts scientists just as much as the rest of us. Most of us assume that others are biased, but that we aren't. As a consequence, many scientists may assume that they are largely free of confirmation bias and don't need to be concerned about its adverse impact on their analyses and interpretation of data.

What should we do to solve these problems?

The open science movement is a great start. Preregistration of one's hypotheses and analyses, especially when done rigorously, is an essential safeguard against not only fraud, but self-deception. Obviously, greater transparency, including providing open data whenever practically and ethically feasible, is also enormously helpful, as it allows well-intentioned scientists to double-check their colleagues' work and conclusions.

Do you think that open science movement and campaigns like the Reproducibility Project and other similar initiatives are sufficient?

They aren't sufficient, but they are and excellent start, and Brian Nosek of the University of Virginia and many others should be applauded for their courageous efforts. Still, we are going to need a large-scale change in academic culture. Department chairs and administrators are going to need to find ways of better rewarding high-quality, dependable science rather than grant dollars. Many of them don't seem to realize that the relentless pursuit of the grant dollar is a virtual recipe for confirmation bias and attendant questionable research practices, such as selective reporting of findings. It's hard to get one's grant renewed if one comes up empty-handed in terms of one's initial findings.

Exactly two months before our conversation, Many Labs Two results have been published. Half of the 28 influential psychological researches were replicated successfully. Is it much or few? Is the glass half full or half empty?

It's hard to say, but I don't think that a 50% replication rate is all that encouraging. We can certainly do much better, and we are beginning to see that in some domains of psychology, including personality psychology, the replication rates appear to be considerably higher. That said, we should be reluctant to conclude that this result means that only 50% of the original findings are genuine. Some of the replication failures may reflect subtle, undetected moderators, although of course to the extent that the moderators are subtle, it may well call the robustness and generality of the findings into question.

There is one more problem plaguing contemporary psychology, and which is not addressed in *Psychological Science Under Scrutiny*. I mean the shift from direct observation of behavior, widely

regarded as an advance in the development of scientific methodology, to introspection. This was demonstrated in an outstanding 2007 **article by Baumeister and his collaborators, and recently confirmed by Doliński** (2018)**, who replicated Baumeister's investigations. Both articles show that over the last few decades, studies of behavior have become a rarity among psychologists. This issue is brought into sharper relief by the fact that the first of the two articles was published in the middle of the decade that the APA announced with great pomp as the decade of behavior. What are your thoughts on this issue?**

I am torn. I'm certainly sympathetic to the assertion that we don't spend nearly enough time observing and analyzing actual behavior, and I'm guilty of this trend too. But things are probably more complicated than many authors imply. For one thing, studying actual behavior is awfully hard. It takes a lot of time and effort, which is certainly one reason why we don't do it enough. But for another, studying actual behavior often contributes to largely unappreciated difficulties with replicability. We learned, or should have learned—from debates in the 1960s and 1970s involving Walter Mischel, Jack Block, Seymour Epstein and others—that studies based on isolated (unaggregated) behaviors are frequently erratic and unreplicable. If we are going to return to studying actual behavior, which I very much support, we are going to need to do it well. Ironically, the replication rates in personality psychology, which relies largely on self-reports, appear to be considerably higher than in much of social psychology, in which we often study behavior. That's perhaps in part because self-reports, as fallible as they are, at least aggregate behavior across multiple imperfect indicators, thereby minimizing errors of measurement. Until recently, many social psychologists did not adequately appreciate the hazards of excessive or exclusive reliance on unaggregated single behavioral indices, such as the speed with which one walks down a hallway after being primed with words connoting old age. I don't share the view that we should be diminishing the value of self-reports of personality traits and other dispositions. Ample data show that these measures yield robust relations with important life indices, including work performance, longevity, happiness, risk for mental illness and the like.

How is it that in spite of information appearing about the crisis and the general decline in trust toward psychology the subject seems to be attracting ever greater numbers of people? As a major, it is shattering records among students in both Europe and America. Isn't this in some way reminiscent of practitioners of a religion who don't pay heed to objective information they are receiving?

I don't see it this way. I see much of this popular interest as reflecting a recognition that psychology continues to address crucial issues that affect people's lives, such as biases, finance, romance, friendships, work, memory, learning, healthy and unhealthy emotions, sleep, eating and the like. Psychology has interesting and important things to say about all of these topics.

Some people claim that psychology attracts people who like fuzzy disciplines, where there are no precise claims and hard knowledge, there is a lot of room for interpretation and no responsibility is required. What is your opinion about it?

There's perhaps some truth to this perspective, although I suspect that it's a bit more complicated than that. Instead, I suspect that psychology tends to attract people who are interested in complex conceptual issues relevant to human nature, and it so happens that many of these issues are often murky and difficult to investigate. Still, I think that there is far too much tolerance for unclear thinking in psychology. Psychiatrist Samuel Guze once made the point that in domains in which we dealing with fuzzy concepts, such as psychiatric diagnosis, we often tend to tolerate equally fuzzy thinking. As he observed, it's precisely in these domains that we instead need to be exceptionally rigorous in our thinking.

In talking about psychiatric diagnosis, we have entered a field that is your original area of expertise. Taking a broad look at psychology as a science and at its applied sub-disciplines, I get the impression that clinical psychology and psychotherapy in particular are not a direct product of the application of scientific discoveries in clinical practice. I also get the impression that these two fields have developed independently of each other, which naturally has serious consequences for practice. What is your opinion on this subject?

It depends. Much of behavior therapy drew directly upon basic animal, and later human, laboratory findings on learning processes. Over time, though, I agree that these two fields have become increasingly disconnected, and that few of the innovations in psychotherapy derive from well-replicated discoveries in basic psychological science. Fortunately, there are some notable exceptions; Michelle Craske et al.'s (2014) recent work on applying basic clinical science research on inhibitory learning to the treatment of anxiety disorders comes to mind as an exemplary example.

Which of the issues in clinical psychology and psychotherapy do you consider particularly dangerous, and why?

Perhaps the biggest threat I see is a resistance to science on the part of a subset of individuals in clinical psychology. Despite the writings of Meehl, Kahneman, Dawes, Faust, Garb and others, some practitioners continue to believe that they can rely largely on their clinical intuitions and clinical experience to inform their choice of assessment instruments and clinical interventions. Needless to say, the histories of medicine, including psychiatry, and psychology tell us otherwise, and warn us of the hazards of this kind of epistemic hubris. In my view, all practitioners should be well aware of the history of disastrous errors in clinical practice, such as Henry Cotton's practices of removal of surgical organs to treat psychosis, prefrontal lobotomy, insulin coma therapy, recovered memory therapy, facilitated communication for autism spectrum disorder and so on. All of these interventions and many more were widely assumed to be effective, even though they actually caused untold harm. As a field, we need to do a better job of (a) selecting students who are scientifically minded and (b) training them in core principles of scientific thinking, which entail the use of systematic safeguards against biases to which we are all prone.

We can judge the power of science by the questions it poses and the answers it provides to them. What big questions have psychology found an answer to so far?

Far fewer than we might like. But in my lifetime, I'd list among the most important discoveries as the following: (a) the realization that genetic factors contribute to virtually all individual differences relevant to human behavior, (b) the realization that the most important environmental factors that contribute to human behavior are nonshared within families, not shared, as presumed by many major models of personality development (e.g., psychoanalytic and social learning models), (c) the realization that essentially all mental disorders are exceedingly multiply determined, and that any model that posits a single gene or specific environmental etiology is likely to be wildly oversimplified, and (d) the realization that although we humans are basically rational, our thinking processes can be distorted by a host of biases of which are largely unaware.

Which big problems still remain mysterious?

More than I can count. But the hard problem of consciousness—how we experience our subjective worlds—remains utterly mysterious, at least to me. Also, we remain a long way from understanding how genes and environment combine or interact to produce human diversity. Even when it comes to genes, a multiplicity of puzzling questions remain. We know that most human individual differences, such as personality and general intelligence, are moderately to highly heritable, and yet we've made only painfully slow progress in identifying specific genes for these phenotypes. Why? It seems likely that they are far more heterogeneous than most of us had assumed. If so, we really have our work cut out for us in identifying their causes.

Will the rapid development of AI help us in finding answers to these questions?

I suspect it will help a bit, but I doubt that it will be anything close to a panacea. Combining extant data in more sophisticated ways, as in machine learning, is indeed helpful. But ultimately, we are going to need more powerful theories.

Would you say that psychologists are not enthusiastic about using self-learning algorithms in their work?

I've actually seen quite a bit of enthusiasm, at least among psychologists who are methodologically sophisticated. If anything, I worry a bit that they are being prematurely overhyped.

Our conversation will also probably be read by young people just getting their careers started, and who are wondering what specialization to explore. What areas of investigation in psychology do you think are particularly promising and would recommend?

I'm reluctant to hazard a guess about which domains of psychology will prove to be especially promising. But were I entering this field again as a young scholar, I'd probably be most interested in the question of the sources of human irrationality. I agree with psychologists Keith Stanovich (2009) and Robert Sternberg (2008) that we've put far too much emphasis on general/analytic intelligence—important as it is—and not nearly enough on such dispositions as scientific thinking, wisdom and intellectual humility, all of which may buffer against pseudoscientific and otherwise irrational beliefs (incidentally, I do not share the view of some psychologists that most or all identified human "biases" are merely artifacts of clever laboratory experiments dreamed up by psychologists—these psychologists need to get out more, and to watch more news and read more world history). I'd also encourage them to explore debiasing approaches—how we can either minimize biases or at least temper their impact.

Who would you point them to as an example to follow?

Daniel Kahneman. Paul Meehl. Carl Sagan. Many others.

What should they be careful of? Can you give them any other advice?

My major piece of advice is to not get too drawn in by intellectual fads. What is "hot" today is likely to be "cold" tomorrow. My other piece of advice is to learn the history of the field. Doing so can be a potent prescription for intellectual humility, as it can remind one of how many smart and highly confident people went very badly wrong.

Let's finish by abandoning our skepticism and intellectual cautiousness and giving our imagination free reign. I'd like to hear about your vision of our field of science for the next two or three decades. I am particularly interested in your response to the question of whether psychology will remain a science that generates a great deal of sexy results that nevertheless do not contribute to forming some general theory, or will it also do the heavy lifting of integrating what we do know, if not into a general theory of human

behavior, then at least into theories of slightly greater scope than those presently existing?

I've learned over the years to be reluctant to prognosticate, largely because I've usually been incorrect in my forecasts. But I do hope and suspect that in the coming decades, we will begin to see greater theoretical synthesis and integration in at least some domains. For example, in the domain of psychopathology, we are gradually beginning to see the rudiments of a coherent theoretical model of mental illness rooted in a relatively small number of biologically influenced personality dispositions that interact with developmental and environmental factors. At the very least, we are also beginning to see the emergence of a recognition that our long-standing models of psychiatric classification, most of which have been premised on the neo-Kraepelinian view of discrete disorders generated by a specific etiology, are likely to be insufficient for many or most psychological conditions.

I have surprise question for you if you agree.

Sure.

I have a list of 30 questions that my readers would like to ask the most influential psychologists in the world. Please draw one.

I'll pick my lucky number 23.

If you had a power to establish one rule for all scientists. What would be this rule?

Preregister your hypotheses and analysis plans, and be clear up front about which of your analyses are planned (confirmatory) and which are unplanned (exploratory). It's the best antidote available against confirmation bias and intellectual hubris, and it can be a humbling experience as an investigator.

Selected Readings

Arkowitz, H., & Lilienfeld, S. O. (2017). *Facts and fictions in mental health*. New York: Wiley.

Heinzen, T., Lilienfeld, S. O., & Nolan, S. (2015). *The horse that won't go away: Clever Hans, clever hands, and critical thinking in psychology*. New York: Worth.

Lilienfeld, S. O. (1995). *Seeing both sides: Classic controversies in abnormal psychology*. Pacific Grove, CA: Brooks/Cole.

Lilienfeld, S. O. (2010). Can psychology become a science? *Personality and Individual Differences, 49,* 281–288.

Lilienfeld, S. O., Ammirati, R., & Landfield, K. (2009). Giving debiasing away: Can psychological research on correcting cognitive errors promote human welfare? *Perspectives on Psychological Science, 4,* 390–398.

Lilienfeld, S. O., Lynn, S. J., & Lohr, J. M. (2013). *Science and pseudoscience in clinical psychology* (2nd ed.). New York: Guilford.

Lilienfeld, S. O., Lynn, S. J., & Namy, L. (2018). *Psychology: From inquiry to understanding* (4th ed.). Boston, MA: Pearson.

Lilienfeld, S. O., & O'Donohue, W. T. (Eds.). (2007). *The great ideas of clinical science: 17 principles that all mental health professionals should understand*. New York: Routledge.

Lilienfeld, S. O., Watts, A. L., & Smith, S. F. (2015). Successful psychopathy: A scientific status report. *Current Directions in Psychological Science, 24,* 298–303.

Skeem, J. L., Douglas, K. S., & Lilienfeld, S. O. (2009). *Psychological science in the courtroom: Consensus and controversy*. New York: Guilford.

Skeem, J. L., Polaschek, D. L., Patrick, C. J., & Lilienfeld, S. O. (2011). Psychopathic personality: Bridging the gap between scientific evidence and public policy. *Psychological Science in the Public Interest, 12,* 95–162.

Wood, J. M., Nezworski, M. T., Lilienfeld, S. O., & Garb, H. N. (2003). *What's wrong with the Rorschach? Science confronts the controversial inkblot test*. New York: Jossey-Bass.

References

Baumeister, R. F., Vohs, K. D., & Funder, D. C. (2007). Psychology as the science of self-reports and finger movements: Whatever happened to actual behavior? *Perspectives on Psychological Science, 2*(4), 396–403.

Cohen, J. (1962). The statistical power of abnormal-social psychological research. *Journal of Abnormal and Social Psychology, 65*(3), 145–153.

Craske, M. G., Treanor, M., Conway, C. C., Zbozinek, T., & Vervliet, B. (2014). Maximizing exposure therapy: An inhibitory learning approach. *Behaviour Research and Therapy, 58,* 10–23.

Doliński, D. (2018). Is psychology still a science of behaviour? *Social Psychological Bulletin, 13*(2). Retrieved from https://doi.org/10.5964/spb.v13i2.25025.

Lilienfeld, S. O., Lynn, S. J., Ruscio, J., & Beyerstein, B. J. (2010). *50 great myths of popular psychology: Shattering widespread misconceptions about human behavior*. New York: Wiley-Blackwell.

Lilienfeld, S. O., & Waldman, I. D. (2017). *Psychological science under scrutiny: Recent challenges and proposed solutions*. New York: Wiley.

Pronin, E., Lin, D. Y., & Ross, L. (2002). The bias blind spot: Perceptions of bias in self versus others. *Personality and Social Psychology Bulletin, 28*(3), 369–381.

Satel, S., & Lilienfeld, S. O. (2013). *Brainwashed: The seductive appeal of mindless neuroscience*. New York: Basic Books.

Stanovich, K. E. (2009). *What intelligence tests miss: The psychology of rational thought*. New Haven, CT: Yale University Press.

Sternberg, R. J. (Ed.). (2008). *Why smart people can be so stupid*. New Haven, CT: Yale University Press.

Wolins, L. (1962). Responsibility for raw data. *American Psychologist, 17*(9), 657–658.

6

Robert J. Sternberg: Intelligence, Love, Creativity and Wisdom

Many scientists live in a fairy world while the world self-destructs. They play a nice tune to themselves and their audience, but the ship on which they are playing is sinking rapidly and they seem either not to care or not to notice.

Robert J. Sternberg

T. Witkowski, *Shaping Psychology*,
https://doi.org/10.1007/978-3-030-50003-0_6

The story of Robert Sternberg's childhood evokes that of the most famous Greek orator, Demosthenes, who forged his abilities through a struggle with his own weaknesses and barriers. To overcome a speech impediment, he practiced diction by placing stones in his mouth, and he enhanced the strength of his voice by shouting over the waves of the sea. The young Robert suffered from what was practically a panicky test anxiety. The IQ-test scores he received were taken by his teachers as an unequivocal sign of his lack of intelligence, and he would probably have come to terms with that label had he not met Mrs. Alexa—a teacher who cared about more than just the results of frequent intelligence tests and expected far more from her pupil than other teachers did. Young Robert eventually cured himself of the test anxiety when, as a sixth grader, he was sent back to a fifth-grade classroom to retake the fifth-grade intelligence test. At that point, he was in the final grade of elementary school, and like all other sixth graders he viewed himself as eminently superior by virtue of his age to the younger kids, whom he perceived as little more than infants. Certainly, he could compete with fifth graders! So he took the test without anxiety, and the test anxiety disappeared and never returned. (That said, he *still* is anxious today when he performs the cello publicly! Apparently, the reduction in anxiety was domain-specific!) But what remained of those experiences was an exceptional interest for a child of that age in the nature of intelligence, how it develops, and how it is measured.

Sternberg is perhaps best known for his research on intelligence and creativity. He is the originator of the triarchic theory of successful intelligence. The theory consists of three main elements: analytical intelligence, creative intelligence and practical intelligence, where analytical intelligence refers to abstract problem-solving abilities, creative intelligence means making use of prior knowledge to deal with new tasks and situations, and practical intelligence is all about the ability to adapt to a changing world (Sternberg 1985a). In its current incarnation, the theory also includes wisdom, or one's skill in using one's abilities to strive toward a common good and make the world a positive, meaningful and possibly enduring difference to the world (Sternberg 2020).

It is precisely this particular conception that brought him recognition, but also led to the acquisition of numerous opponents. By introducing into his theory the notion of practical intelligence, he took up the mantle of Edward Thorndike, who had already proposed in the 1920s including social intelligence in measures of intelligence. But social intelligence proved elusive to gauge and to display in numbers. The world was fixated on academic intelligence, easily measured using IQ, and forgot about the ideas advanced by Thorndike. Sternberg disrupted the calm of many intelligence researchers who were accustomed to old psychometric instruments and pretty numbers, but which had little in common with reality. And while his conception found a permanent place in psychology textbooks, it remains used by only a relatively small number of people in the field of education.

As for Sternberg, he is a quite typical example of the stereotypical observation that the majority of psychology students select that field of study in order to solve their own problems. He himself admits that he has always solved his problems through psychology, but usually by creating his own theories rather than by using other people's. As he puts it tersely: "I do my best when the tune I play is the one I write" (Sternberg 1985a).

One example that serves to illustrate this approach is his adventure with the psychology of love. He was at a point in his life once in which an intimate relationship wasn't going as he had hoped it would, but he couldn't quite figure out why. He did some reading on the psychology of love, but it failed to adequately address the problems he was having. And this was the starting point for his triangular theory of love, which also brought him fame, not so much fortune, and … plenty of enemies. According to this theory, commitment, passion and intimacy are the three main components of love. The combination of these three elements in different ways results in different types of love (Sternberg 1986).

Sternberg's contributions to the world of psychology are numerous. Apart from the psychology of love, he has also conducted extensive research into hate, higher mental functions, including wisdom, creativity, styles of thinking, cognitive modifiability and leadership. His impact in all of these areas is clear and significant.

In contrast to the trend dominant in contemporary science, Sternberg is a vocal advocate of taking science out of laboratories and of studying human behaviors in people's natural environment. He has prioritized studies in schools and in everyday life over those performed in isolated conditions, and he eagerly takes up studies encompassing multiple cultures.

Professor Sternberg, our conversation is taking place at the same time as the debut of a book which you are the editor of, *My Biggest Research Mistake: Adventures and Misadventures in Psychological Research* (Sternberg 2019b). This is an unusual book, considering that people in general, and scholars in particular, don't like discussing their mistakes. Was it difficult to convince your contributors to participate in the undertaking?

No. Almost everyone I invited to write agreed to contribute. I was surprised. And then, there were some people who were too busy and others who probably, understandably, would have liked to see their mistakes remain deeply buried.

In your opinion, do the reflections contained in the book explain the causes of the crisis in psychology that is increasingly being discussed in an entirely open manner?

Different scholars see different crises. I realize many scholars believe replicability to be the biggest crisis. Certainly, it is an issue for the field to deal with. In my view, however, the biggest crisis in psychology is that people increasingly are focusing more and more carefully on smaller and smaller problems that are less and less relevant to people's everyday lives. The open science movement has many pluses, especially transparency, but *all* movements, including this one, have their minuses. I believe a potential minus is that researchers will be spending increasing amounts of time on (a) documentation, (b) assuring, reassuring and re-reassuring that their results are replicable to the point that they wait too long to publish, if they ever publish and (c) verification as opposed to innovation because they are afraid that they will be attacked, possibly savagely and personally, for lack of replicable or otherwise verifiable results. That is not to say that replication is not a problem. But sometimes the solutions we generate have collateral effects we fail to anticipate. I think many scientists may be afraid to be creative because creativity involves risk and taking risks these days leaves one potentially exposed to attack on a massive scale that was not possible before the rise of the Internet and social media. Scientists like to view themselves as independent and as creative, but they are as susceptible to following the crowd and even joining Internet mobs as anyone else. We all like to think we are special but then discover that we all have the same human foibles and tribal instincts as everyone else. This is a point of my triangular theory of creativity. Creativity requires us to defy the crowd—which few scientists do—to defy ourselves—which scientists also are loath to do once they think they have a formula for success—and to defy the Zeitgeist or prevailing worldview—which few scientists even contemplate. In my edited book, *Psychologists Defying the Crowd* (Sternberg 1985b), one learns that often some of the greatest opposition creative scientists have faced is not from laypeople or even politicians but rather from other scientists. None of these attacks on creative individuals are anything new in the world. We all know what happened to Socrates—Athenians turned on him and forced him to drink hemlock. Many of Jesus' disciples turned on him, sooner or later. In science, much later, Ignaz Semmelweis fared no better—his medical colleagues turned on him when he told them to wash their hands. Todd Lubart and I in our book on *Defying the Crowd* give many examples of how people turn on creative individuals (Sternberg and Lubart 1995). As I have pointed out in many writings, being creative is hardest not because of people's lack of creative abilities but rather because people don't want to pay the price for creativity, which often is rejection and ostracism. And it is much easier to attack others—to reject them, to ostracize them—than to come up with your own creative ideas. That said, there is room in science and elsewhere for many different types of scholars. We need some people who specialize in coming up with ideas and it probably helps to have others who specialize in critiquing these ideas. What has changed in recent times, perhaps, is the reward system, which today seems to favor the latter over the former.

I wouldn't like for your words to be misunderstood or misused in a manner contrary to your intentions. When I engage in discussion with people promoting pseudoscience, they frequently invoke the example of Semmelweis and other scientists who displayed courage in defying the crowd and paid a high price for it. They are convinced that scientists do not accept their claims in the same manner as they ignored Semmelweis and many others. For them, the opinions of scientists tell us little, apart from showing them to be narrow

minded people. So how, if not with studies and replications, are we able today, in the midst of a flood of pseudoscientific claims, able to distinguish creative ideas from insanity and plain fraud?

I do not think I said anywhere that I oppose or even am unfavorable to replications. Replications are today and always have been useful. What I did say is that it is important to achieve a balance. I am not sure we have achieved that. So, to be clear: I support replications and have done them myself. Indeed, I am just about to submit a coauthored paper that is a replication and extension of my own earlier work on intelligence—scientific reasoning, in particular. Also, I am a coauthor on a submitted paper that is a replication and extension of my work on love cross-culturally. So, I do replications myself! But the reward system of science and society needs to reward the risks people take when they are creative and wise, and if creative people are put down—going back at least to Socrates but to so many others, including Jesus—we are not going to achieve the balance we need. Sure, none of us is Socrates or Jesus. But too often, people who try to say something new get shouted down for saying it. Some of what they say is inevitably going to be wrong. We need people to point that out. But we also need people to recognize contributions that are off the beaten track.

What I did say is that I do not view replication as the single biggest problem facing psychological science. I said that lack of creativity is, because there is so much pressure to follow the crowd. There is more pressure today because of social media. I first spoke of the importance to creativity of defying the crowd in a coauthored paper with Todd Lubart back in 1991 so there is nothing new in my saying it. People, including scientists, are afraid to defy the crowd. Scientists join crowds as much as does anyone else. One of my mentors, Wendell Garner, told me that he had been told by Michael Posner that the hardest papers to get accepted are crappy ones because they are crappy and highly creative ones because they are highly creative. Endel Tulving and Stephen Madigan (1970) made a similar point in a paper in the *Annual Review of Psychology*—that there were very few papers published that they wish they would have thought of themselves. That was true when Garner, Posner, and Tulving talked about it and it is true today. Thomas Kuhn made a similar point. Defy the existing paradigms and the rewards are much more likely to be long-term than short-term. And in the short term, you will have large numbers of colleagues telling you what a fool you are. In the field of intelligence, for example, IQ tests are very popular—just as they were a century ago. That's not exactly defying the crowd. We also show, as a field, a lack of wisdom. There is too little work on how psychology can be directed toward achieving a common good. All that said, fraud is always awful, and fraudulent scientific theories and findings need to be uncovered and exposed! Frauds, like merciless critics of creativity, always will be with us.

You said before that the system favors people who specialize in critiquing ideas. Meanwhile, in science, and particularly in psychology, we are continually confronted with the immortality, or at least the inexplicable longevity of some ideas—for example, "academic intelligence" and means of testing it, an idea you have harshly criticized. Is their durability not a result of insufficient criticism?

There are some ideas that way outlast their usefulness, if they ever had usefulness. However, the idea of "academic intelligence" is not one of them. Academic intelligence, *g*, or whatever you want to call it, is an idea that was useful in the past and still is useful. Scores on measures of the *g* factor correlate weakly to moderately with many outcomes in life, not just academic ones. The problem is not with the concept of *g* but rather with the overuse of and even obsession with it. Because the concept has had some success—as well it should have—researchers in the field of intelligence have not been particularly bold in going beyond it, unless you look at hierarchical models, the first of which was proposed by—of all people—Cyril Burt in 1940—as a giant step forward. My argument has not been against academic intelligence. Quite the contrary—it is part of my own theory. My argument is that we make far too much of it. As I said in my 2019 invited address at Society for Research in Child Develpment and will write in a forthcoming book on adaptive intelligence, IQ's rose 30 points in the twentieth century (the so-called Flynn effect, see Flynn 1987) and what do we have to show for it? Every year, so far, carbon emissions have been increasing, global climate change is getting worse, air pollution is killing countless people, we are making ever-more powerful explosives, and more people are dying from hunger and disease (see Sternberg 2019a).

Income disparity is increasing in many countries, to the point that it is unsustainable. If that is what 30 points of IQ has bought us, something is missing, I think. Intelligence is the ability to adapt to the environment—or at least that is the essence of its definition over time—but we, as a species, are doing a pretty awful job of adapting to the environment. Rather, we are destroying it and ourselves in the process. That is why I have argued that people need to show more "successful intelligence," which includes creativity, common sense, and wisdom in addition to analytical skills. How many leaders in the world today, in any domain, are showing much of any of these attributes? Certainly not many in my country, where so many of them go to "prestigious universities" in part because of their high IQs. Rising IQ is not going to save humanity (*Homo sapiens*, to be exact) from being the first species to extinguish itself and take countless other species with it.

Let's go back to your book *My Biggest Research Mistake*. As its editor you perhaps enjoy greater knowledge than anyone else concerning the pitfalls a researcher can fall into. What should we pay the most attention to in order to begin speaking about the crisis of psychology in the past tense?

We should find a better balance in the creative, analytical, practical and wise aspects of the scientific endeavor. The creative aspect is involved with generating novel and potentially useful ideas. The analytical aspect is involved with making sure our ideas are sound and correctly tested. The practical aspect is involved with the applicability of results to the world. And the wisdom-based aspects are involved with whether our research ultimately will help achieve a common good. At present, the analytical element is getting the lion's share of the attention, which follows from increasing selectivity of scientists for their analytical skills and attitudes but not necessarily for their creative, practical and wisdom-based skills and attitudes. All these aspects are important but they need to be balanced. Using memory and general analytical tests to admit students to graduate school is a sketchy idea. My recent research with my wife Karin Sternberg has shown that these tests show little or sometimes even negative correlations with tests of scientific reasoning, such as formulating alternative hypotheses, generating experimental ideas, and drawing conclusions (Sternberg and Sternberg 2017; Sternberg et al. 2017, 2019).

We have created an atmosphere whereby many people are afraid to be truly creative. They are understandably afraid that, if they do anything highly creative, they will find themselves (a) out of paradigm, (b) with rejected articles and grant proposals and (c) ignored or shunned by their colleagues. The greatest creative rewards are for those who are forward incrementers—who stay one step ahead of others on whatever path the others are taking. As Sternberg et al. (2002) argued in our book on a propulsion theory of creative contributions, you need a lot of courage either to try to take giant steps forward (advance forward incrementation, as we called it), to redirect a field, or to try to reinitiate a field—to start it over from scratch. So, scholars become more and more careful about smaller and smaller stuff because of the current costs of making even small mistakes. Once upon a time, people made mistakes, apologized, learned from the mistakes, and then tried not to make the same mistakes again. Today, social-media mobs are ready to belittle people even for admitting to error. The Trumpian mentality that has become increasingly prevalent in politics is becoming increasingly prevalent in science too—even among those who would never consider voting for Trump and would think of themselves as having nothing in common with him.

Could you say a few more words about what you describe with the phrase "Trumpian mentality"? I am counting on this book being read also by people for whom Trump and everything associated with him will be nothing more than history. This term might not be entirely clear for them.

Donald Trump is almost certainly the worst president the USA has ever had (another questionable correlate of rising IQs?). I don't know what his IQ is—but in terms of wisdom, he simply does not have it. He does not know what he doesn't know, and worse, doesn't care. He apparently does not read. He does not expose himself to alternative points of view. He has strong opinions on things he knows nothing about. He is a racist and a xenophobe. He hires and fires many people, for the most part keeping on only the most incompetent and malign. He is amoral. These are not qualities one would want in any leader. But he is not the problem. The problem is that, despite the rising IQs in the twentieth century, people in more and more countries are electing malign, toxic leaders like him. It's a worldwide phenomenon, not just a US one. This is seriously awful, because by any standard, worldwide democracy is on the decline (obviously, this idea is not in any way original with me—there are now multiple scholarly books about it—for example, Levitsky and Ziblatt 2018; Mounck 2018). The problem is not just with the leaders, but that many people outside of politics are adopting his style—loudmouthed, uninformed, brutal and belittling. Social media almost certainly are partly at fault, but in the end, we must take responsibility as humans for our own shortcomings. This kind of *ad hominem* attacking has made its way into science, as five or fewer minutes on the Internet will reveal to anyone who troubles to look for it. The Trumpian mode is to advance oneself by belittling and if possible destroying others instead of by coming up with one's own creative, analytically sound, common-sense, wise ideas. That modus operandi is to be found in all walks of life today, including but certainly not limited to science.

You have long been a harsh critic of what is taking place in psychology. In the first two sentences of the *Preface* to the aforementioned earlier book *Psychologists Defying the Crowd,* published in 2002, you wrote: "Today, if you are a graduate student on the job market who has training and published research in cognitive-neuroscientific approaches to the study of memory, and you have half a brain, you will have several good job offers and maybe quite a few such offers. If, however, you are a graduate student combining techniques of personality research with those of cognitive research to study creativity, your job pickings may be slim or nonexistent, even with a whole brain (and a good one at that)." This is a sort of diagnosis of our entire field. Has anything changed in the years since you wrote that? For the better, or rather for the worse?

I am totally in favor of neuroscientific approaches. But I am not in favor of any one approach as "better" than other valid approaches. As I have said before, the problem with psychology today, as in the past, is its susceptibility to the fads of the moment. At different times, different approaches are supposed somehow to be panaceas—the psychodynamic approach in clinical psychology, or the behavioral and then the cognitive approach in experimental psychology, now the neuropsychological approach. Different approaches address different kinds of questions.

There is no one right or best approach. It depends on the problem you want to solve. When you monogamously marry yourself to an approach, as so many investigators do, you close off the possibility of answering questions that approach does not well address. So, your paradigm creates the walls that limit the kinds of questions you allow yourself to try to answer. I should note that I have been, in the early days of my career, as guilty as anyone else of believing that one approach (in my case, the cognitive one) held all the answers. Of course, I was wrong. One never gets over being wrong, at least some of the time!

I am not skeptical of neuroscience. I am skeptical of the view that any one approach is a panacea. I have said the same before of behaviorism when it was riding high, and then of cognitivism when it was riding high, and of factor analysis, and so on. So, one might conclude I am skeptical of all approaches, but only when people blindly follow them like a religion. When it comes to blind faith in paradigms, scientists can be much like nonscientists.

These are not the only problems that you perceive in psychology. You mentioned in your talk at the Wisdom Research Forum 2016 that a lot of research "...is analytically strong, but doesn't matter." Why, in your opinion, are so many psychologists firing such large methodological cannons at such insignificant targets?

Because that is what many researchers want, what journals publish, and what granting agencies fund. In the past, more researchers were willing to defy the crowd and go their own way. But the past pressures of hiring, tenure and promotion, negative attention from social media, and sheer competition in the field were not what they are today. We have more people today chasing fewer and fewer rewards (and indeed, grant funding has declined significantly in many countries), leading researchers to be conservative because they understandably fear for their careers.

In the same lecture you gave examples of civilizational achievements such as the development of technology that we can use to exterminate ourselves, climate disaster, unequal health distribution and many more, which are the utter opposite of the optimistic vision of the world presented by Steven Pinker in his last two books (Pinker 2011, 2018) **so enthusiastically received by Western civilization. I take it you do not share that enthusiasm?**

I do not share Pinker's enthusiasm. I think he is too caught up in averages and looking at the short-term future. If humanity extinguishes itself, as it seems determined to do, through global climate change, weapons of mass destruction, pollution, antibiotic-resistant bacteria, terrorism, idiotic politics such as we have in the USA and elsewhere, or whatever means it chooses, all those nice graphs won't mean a whole lot to the cockroaches and ants that are left on the Earth. For people in the upper middle and upper classes like Pinker, things are great in the short run; for most of the rest of the world, things could be a heck of a lot better. Things really have improved enormously for the upper classes. Pinker acknowledges many of these and other blips. But Pinker and his colleagues perhaps should spend more time in inner-city slums or vast areas of rural poverty. Average life span in Pinker's own country, the USA, is going down as a result of illicit-drug deaths. We may be, therefore, at an inflection point, where things will start getting worse. And they will be *much* worse for our children and grandchildren, who will have to live with the global warming and air and water pollution we are leaving to them. In my own country, we also are experiencing unprecedented levels of corruption, together with legal and political systems that seem powerless to do anything about it. That should tell us something. The world wars had unprecedented levels of deaths and, should there be another world war, the death toll likely would be staggering. When I look at the rise of autocrats around the world, including in my own country, it becomes much easier to understand how little people learn from history and how not only susceptible they are, but eager they are to repeat it.

This is probably also true for psychologists. From time to time, they engage in ethically questionable activities just as they did at the beginning of the twenty-first century, when psychologists affiliated with the APA were involved in work on techniques for interrogations to be applied as part of the war on terror (Risen 2014). And this is not the first time that people from our discipline took active part in researching and developing methods used against other people. We can recall the example of the CIA-inspired MKUltra program. How do you, as a past president of the American Psychological Association, judge the participation of psychologists in perfecting torture

techniques, which is probably a fair phrase to use in referring to these "interrogation methods"?

That is a complex matter. APA totally blew it in terms of how they handled the whole affair, but that is a story for another day. I now have quit APA, despite being a past-president. Obviously, those individuals who engineered torture gave psychology an awful name and deserve the criticism they got. That said, they are a small minority of psychologists. They should have known, and probably deep down, did know better. But the field has done many bad things over the years—some have participated in torture, others in harmful treatments for people with clinical challenges, others in misuse of standardized tests in schools. We all could do better, myself included. That is why I emphasize wisdom in my work—using one's knowledge and abilities to obtain a common good and also why I emphasize positive creativity—being creative in ways that make the world a better place.

How is it that in spite of all these problems we've discussed, and which psychology must handle, along with the general decline in trust toward psychology, the subject is attracting ever increasing numbers of people? As a major, it is shattering records among students in both Europe and America. What are the reasons for this state of affairs?

That's great and is as it should be. Let's hope that the next generation of psychologists views their job as one of making the world a better place to live—that they make a positive, meaningful, and enduring contribution to the world, at any level.

Your scientific interests are quite broad, but in many statements you emphasize that your primary mission is to change the way people think about intelligence and means of measuring it. Looking back, could you say that you have at least partially succeeded in that mission?

No. I've been a serious failure. When I started as an assistant professor at age 25, I was very optimistic that somehow I would be able to change the world. I haven't and in the time I have left I probably won't. I'll hope that my children, my students, your students, or some in the next generation are more successful than I have been.

Merciless self-evaluation for someone who is one of the most frequently cited psychologists in the world, a recipient of the most prestigious awards, and originator of several theories that enjoy broad recognition. Perhaps the cause lies someplace outside you, and something stops people from abandoning their customary and limited way of thinking about intelligence? What are the reasons for this?

I want to make clear that my theories, like everyone else's, are wrong. All scientific theories are wrong. Science exists to keep making theories better. I would have hoped, when I started my career, that at least in the field of intelligence, people would think more broadly by this time. I was wrong. I asked for too much, I suppose. But there are lots of reasons why people stay with traditional theories and measures of intelligence. First, the theories are useful—as far as they go—they just don't go far enough. Second, *g*-based constructs are easy to measure. Third, people are willing to keep doing the same thing. There are some scientists, literally, who have spent almost their entire careers showing over and over again that *g* predicts this, that, and the other thing. Imagine if they said to themselves—"OK, I have shown 350 (or more) times that *g* predicts yet something else. Now, can I find some new constructs that perhaps go beyond *g*, maybe even well beyond it?" Fourth, there is an enormous literature to draw from and build upon, and most scientists understandably want to build on the work of others. Fifth, working on *g*-theory is safe—it's a way to achieve publications, awards, grants, invitations to meetings at testing companies, and the like. Sixth, the leadership of Western societies has for the most part bought into the *g* notion. Because *g* is correlated, for various reasons, with socioeconomic status, it ends up being a way of cloaking continuation of the existing social and economic order in "meritocratic" clothing. But in terms of why people have not bought into my theories, I can scarcely blame others for my own failures in convincing other people of the value of my work. That's my inadequacy and I take responsibility for it. I tried and failed. We need better theories than mine, or more powerful persuasive voices, or both.

Interestingly, we psychologists study the majority of such behaviors which you are talking about, and we tell others that they are

biased in decision making, thinking, and which limit our development. It seems like the shoemaker's children are ill-shod...

The fundamental principle of interpersonal attraction is that we are attracted to others like ourselves. And so, scientists tend to look for others like themselves who are, for the most part, forward incrementers who don't threaten the existing order.

Probably we could spend a lot more time analyzing the problems inherent into our discipline. But let's set them aside for now, to take a closer look at psychology's achievements. Your studies and publications have reached the top of the mountain in terms of citations and are a fundamental part of our field's accomplishments. Which of your discoveries do you consider the most important in the development of science, and why?

None of them are earth-shattering, unfortunately.

I think my most influential theories have been my theory of successful intelligence and my triangular theory of love. The theory of successful intelligence perhaps is of some importance because our narrow conception of intelligence is destroying our species and our world. We use the label "intelligent" to apply to people who are doing destructive things that are harming humanity and many other species as well. In the augmented theory of successful intelligence, wisdom plays a key role.

The triangular theory of love has been of some import because everyone wants to understand love and the theory has had some success in doing so. We are starting a website, lovemultiverse.com, to help people with their love relationships.

That said, none of the theories has really had much impact on the way society goes about doing things, so in a larger sense, nothing I've done has mattered a whole lot.

Is there any particular type of criticism aimed at you that you feel is particularly serious and justified?

There are so many I would not know for sure where to start. No one is as critical of me as I am, to my knowledge. I could not list all the various criticisms I've heard. The one I've heard the most is that my theory of intelligence is too broad. But if you look at the current mess in the world and the fact that IQs rose 30 points in the twentieth century, IQ doesn't seem to provide any protection against any adverse outcomes for the world. If anything, it hastens those adverse outcomes as we keep doing what we can to destroy the very civilization we have created. The most trenchant and important criticism is my own—that I have changed absolutely nothing and thus have failed in my attempts to make the world a better place.

There's an old joke that goes: "How many psychologists does it take to change a light bulb? One, but the light bulb really has to want to change." While, in relation to a light bulb, our dependence on motivating an object to change may seem quite poor, when it comes to an attempt by one psychologist to change the world, the world's readiness to change appears to be of key importance. The world must really have to want to change. But let's get back to your research. It seems that in both your studies on wisdom and on ethical behaviors you are approaching, or perhaps even crossing the boundary of what Hume's guillotine demarcates in science. What is your take on this principle? Should scientists not only explain reality, but also tell us what we "should"?

What good is science exactly if it does not tell us how we can live better? What good, for example, is climate-change research if it doesn't tell us to stop overheating the planet? What good is intelligence research if its practitioners can do no more than show, for the 35,005th time, that yet something else correlates with IQ? Have we no self-respect? That is to me the problem with science as so many scientists see it. We are like the band on the Titanic, playing away furiously while our ship sinks. At the rate we are going, there won't be any science in the future whose boundaries we have to argue about because with all the harms we have created through science—pollution, global climate change, drug-resistant bacteria, weapons of mass destruction—we won't be around to have these philosophical discussions. Many scientists live in a fairy world while the world self-destructs. They play a nice tune to themselves and their audience, but the ship on which they are playing is sinking rapidly and they seem either not to care or not to notice. What good are IQs that rose 30 points in the twentieth century if they lead us to elect and then cheer on corrupt autocrats who do whatever they can to cause the world as we know it to self-destruct? Do we want more high-IQ people who will be even more effective at doing so? Why not try to measure and improve positive creativity, common sense, and wisdom too? Too hard? Too threatening? Not worth the bother?

Let's get back to the achievements of our field. The power of science is found in the questions it asks and the answers it provides. What great questions has psychology managed to answer so far?

Science has answered many questions, psychological science, in particular, not so many. I can't think of any off-hand that psychology has answered definitively.

Which crucial questions are still waiting for answers?

Oh, pretty much everything. Why do we insist on being self-destructive? Why do we elect dictators and then rejoice while they corruptly and even flagrantly pursue their own self-interest at our expense? Why do scientists so much prefer small problems to ones that might actually benefit society? Why do we insist on calling people intelligent who are destructive of themselves and others?

Do you think search for answers to these questions can be aided by the intensive growth of AI that we are witnesses to?

AI can provide many benefits. What it is doing right now is obscuring the search for meaning that characterized past generations. Technology can lead us to more comfortable lives but not necessarily to more meaningful ones.

What are the greatest challenges facing psychology in the twenty-first century?

How to be culturally and globally relevant and positively impactful at a time that science is more concerned with the degrees of freedom in your *F* test and whether you preregistered your hypotheses.

I hope that our conversation will also be read by young people just getting their careers started, and who are wondering what specialization to explore and how to do it. You wrote the whole book with advice for them (Sternberg 2016), so you are the right person to ask you about some advice for them. What areas of investigation in psychology do you think are particularly promising and would recommend?

The area does not matter. What matters is finding important problems with scientific and societal implications and then doing the research that will help make a positive, meaningful and enduring difference to the world, instead of pleasing people who have their noses buried in trivial details. Most of all, find problems about which you are passionate and don't listen to all the noise in the background. If you do what you believe is meaningful, you will get a lot of criticism. Learn to live with it.

Who would you point them to as an example to follow?

I think Carol Dweck is one. Her research has had some real impact on schools and on how people perceive themselves. Roy Baumeister is another. He is incredibly creative, which of course is why so many people want to take him down—he makes them uncomfortable with themselves. That is not to say that all his ideas are perfect—but they are large and creative in a way that is rare today. Marty Seligman has done a lot of good, I think, with his ideas about positive psychology, although many of the ideas now have been trivialized by others in a kind of feel-good psychology. I think David McClelland, with his work on achievement motivation, was really admirable, as was Irving Janis and his work on groupthink. These are people whose research really helped change the world. They concentrated on important problems rather than what paper their next article would be in or who would fund their next grant proposal or what others would say about what they did. Scott Lilienfeld has been amazingly creative: Every paper he writes is one I wonder why no one else thought of. Susan Fiske has done enormously creative work in social psychology. The late Paul Baltes essentially invented the field of wisdom research. He was amazing. And my mentors, Gordon Bower and Endel Tulving, were both hugely innovative.

What should they be careful of? Can you give them any other advice?

We should be careful about joining social-media mobs on the internet—we are showing ourselves not to be much different from some of the political extremists we otherwise claim we detest. We should be careful about following fads just because everyone else is. We should be careful about the analytical component of research dwarfing the creative and wisdom-based ones. We should be careful of research that doesn't matter except to help us get our next publication or grant. We should be careful to focus on big important problems where we really can make a difference to the world and make it a better place.

And what about taking research from the laboratory to everyday life? I know that you are a strong advocate of this, and during our conversation you haven't mentioned anything about it.

I have spent my whole career trying to do that and have largely failed. Not much more to say. I'm 69. There is perhaps a tiny chance I still will succeed. Otherwise, it will be our children and students who are left to change things. But I do believe we are running out of time. Psychologists in my field are busy tracing down the latest correlate of *g* or the latest brain model of intelligence, all of which is fine. But if there is no world left, what difference will it all have made? While we are pursuing these problems, real adaptive intelligence—the kind that might save humanity from itself—is getting little if any study at all. If we are really good at what we do, then we should be trying to save the world before it is too late. But you don't get publications in scientific journals and grants and awards for that, so most scholars in the field understandably go for the short-term rewards. Where is intelligence research, or research in other fields, when we need it to save us from destroying ourselves? But then, why are states in the USA that are seriously being harmed by climate change supporting a president who denies climate change? Go figure. My oldest daughter Sara, a law professor, is doing research that I believe is really valuable, showing how the legal system, which is supposed to be "just," at least in the USA systematically favors the wealthy and well-connected. We need more people in all fields taking on problems such as that one.

How do you imagine our field of science over the next two or three decades?

I have no idea. I can't even predict what will happen tomorrow. Like many others, if I looked at the world two years ago, I never would have imagined the mess it would be in today with the emergence of autocrats in so many different countries, including my own; nor would I have imagined how slavishly people would embrace them.

What are other issues we have not talked about yet but you would like to mention to our readers? Such as a message about contemporary psychology, a big issue or important question about psychology?

One last thing: We all have our individual, collective and varied contributions to make. Not everyone will share my value system. Some will see my value system as wrong or orthogonal to science. Of course, we need people with diverse value systems. But the world really is in trouble. A lot of the land shown on maps today is not really land anymore; it's underwater. The State of Louisiana in the USA, for example, is losing one football fields' worth of land every hour. The State of California is getting unprecedented wild-fires. There is catastrophic flooding in the States of Iowa, Nebraska, and elsewhere. Islands are going under. Really, we need to think of ways to get psychologists to pay more attention to these problems, whether or not those are the problems that will get the most immediate plaudits from their colleagues. Others may disagree, but I think we need to spend more time attacking serious world problems and less time attacking each other. And yes, I realize I may be attacked for that statement—so be it. I'm 69 and won't be around all so long: The new generation has the chance to make a really meaningful positive difference to the world where my generation did not do so, however good their intentions. I am very hopeful of the future if the next generation follows its leaders like Greta Thunberg who recognize they must cure the ills my generation created or failed adequately to address.

Selected Readings

Sternberg, R. J. (1996). *Successful intelligence*. New York: Simon & Schuster.
Sternberg, R. J. (1997). *Thinking styles*. New York: Cambridge University Press.
Sternberg, R. J. (1998). *Love is a story*. New York: Oxford University Press.
Sternberg, R. J. (1999). The theory of successful intelligence. *Review of General Psychology, 3*, 292–316.
Sternberg, R. J. (Ed.). (2002). *Why smart people can be so stupid*. New Haven: Yale University Press.
Sternberg, R. J. (2007). *Wisdom, intelligence, and creativity synthesized*. New York: Cambridge University Press.
Sternberg, R. J. (Ed.). (2020). *Human intelligence: An introduction*. New York: Cambridge University Press.

Sternberg, R. J., & Grigorenko, E. L. (2000). *Teaching for successful intelligence*. Arlington Heights, IL: Skylight Training and Publishing Inc.
Sternberg, R. J., & Spear-Swerling, L. (1996). *Teaching for thinking*. Washington, DC: American Psychological Association.
Sternberg, R. J., & Sternberg, K. (2008). *The nature of hate*. New York: Cambridge University Press.

References

Burt, C. R. (1940). *The factors of the mind*. London: University of London Press.
Flynn, J. R. (1987). Massive IQ gains in 14 nations. *Psychological Bulletin, 101*, 171–191.
Levitsky, S., & Ziblatt, D. (2018). *How democracies die*. New York: Broadway Books.
Mounck, J. (2018). *The people vs. democracy: Why our freedom is in danger and how to save it*. Cambridge, MA: Harvard University Press.
Pinker, S. (2011). *The better angels of our nature: The decline of violence in history and its causes*. New York: Viking.
Pinker, S. (2018). *Enlightenment now: The case for reason, science, humanism, and progress*. New York: Viking.
Risen, J. (2014). *Pay any price: Greed power, and endless war*. Boston: Mariner Books.
Sternberg, R. J. (1985a). *Beyond IQ: A triarchic theory of human intelligence*. New York: Cambridge University Press.
Sternberg, R. J. (Ed.). (1985b). *Psychologists defying the crowd: Stories of those who battled the establishment and won*. Washington, DC: American Psychological Association.
Sternberg, R. J. (1986). A triangular theory of love. *Psychological Review, 93*, 119–135.
Sternberg, R. J. (2016). *Psychology 101½: The unspoken rules for success in academia*. Washington, DC: American Psychological Association.
Sternberg, R. J. (2019a). A theory of adaptive intelligence and its relation to general intelligence. *Journal of Intelligence*. Retrieved from https://doi.org/10.3390/jintelligence7040023.
Sternberg, R. J. (Ed.). (2019b). *My biggest research mistake: Adventures and misadventures in psychological research*. Los Angeles: Sage.

Sternberg, R. J. (2020). The augmented theory of successful intelligence. In R. J. Sternberg (Ed.), *Cambridge handbook of intelligence* (2nd ed., pp. 679–708). New York: Cambridge University Press.

Sternberg, R. J., Kaufman, J. C., & Pretz, J. E. (2002). *The creativity conundrum: A propulsion model of kinds of creative contributions (essays in cognitive psychology).* New York: Psychology Press.

Sternberg, R. J., & Lubart, T. I. (1995). *Defying the crowd: Cultivating creativity in a culture of conformity.* New York: Free Press.

Sternberg, R. J., & Sternberg, K. (2017). Measuring scientific reasoning for graduate admissions in psychology and related disciplines. *Journal of Intelligence, 5*(3), 29.

Sternberg, R. J., Sternberg, K., & Todhunter, R. J. E. (2017). Measuring reasoning about teaching for graduate admissions in psychology and related disciplines. *Journal of Intelligence, 5*(4), 34.

Sternberg, R. J., Wong, C. H., & Sternberg, K. (2019). The relation of tests of scientific reasoning to each other and to tests of fluid intelligence. *Journal of Intelligence, 7*(3), 20. Retrieved from https://doi.org/10.3390/jintelligence7030020.

Tulving, E., & Madigan, S. A. (1970). Memory and verbal learning. *Annual Review of Psychology, 21*(1), 437–484.

7

Robert Plomin: Behavioral Genetics

My religion is science, not individual scientists. I believe in science, but I try to resist the temptation to believe in scientists and to hold them up as heroes.

Robert Plomin

T. Witkowski, *Shaping Psychology*,
https://doi.org/10.1007/978-3-030-50003-0_7

As an adolescent, Robert wondered if he had been adopted. He recounts in his book that he grew up in a one-bedroom flat in inner-city Chicago without books (Plomin 2018). His father worked on the assembly line at a car factory, later becoming a layout engineer. No one in his family went to university, including his parents, his sister and a dozen cousins who lived nearby. However, his parents encouraged him to go to the public library and became an avid reader at an early age, bringing bags of books home from his local public library. He always did well at school, worked hard, was conscientious and he persevered. At that time, he often wondered where his interest in books and school came from, given that his family showed little attachment to these things. He found the explanation when he began studying genetics. As a teenager, he didn't realize that although the first law of genetics is that like begets like, the second law is that like does not beget like. Genetics makes first-degree relatives 50% different as well as 50% similar.

His interest in behavioral genetics developed at graduate school in Texas, then a leading center for the subject. After finishing his Ph.D., he got his dream job at the University of Colorado at Boulder with a joint appointment in the Department of Psychology and the Institute for Behavioral Genetics, the only institute of its kind in the world. He decided to create a long-term longitudinal adoption study of psychological development. From 1986 until 1994 he worked at Pennsylvania State University, studying elderly twins reared separately and twins reared together to study aging, and he currently works at the Institute of Psychiatry of King's College London. Plomin is presently conducting the Twins Early Development Study of all twins born in England from 1994 to 1996, focusing on developmental delays in early childhood, their association with behavioral problems and educational attainment.

Among Plomin's numerous achievements, of particular distinction is the importance of non-shared environment, a term that he coined to refer to the environmental factors that reduce the similarity of individuals raised in the same family environment. Moreover, he has shown that many environmental measures in psychology show genetic influence, and that genetic factors can mediate associations between environmental measures and developmental outcomes.

In 1994, he was one of the 52 signatories of "Mainstream Science on Intelligence," an editorial written by Linda Gottfredson (1994) and published in the *Wall Street Journal*, which declared the consensus of the signing scholars on issues related to intelligence research following the publication of *The Bell Curve* by Richard J. Herrnstein and Charles Murray (1994). This was a very significant event in his life. As he frequently points out, at the time when he began his research on behavioral genetics, it was practically scientific suicide to speak openly about the genetic conditions of individual differences, particularly regarding intelligence. As he emphasizes, for many years he kept his head down with his work. He found the courage to speak up only after *The Bell Curve* was published. Because of the general aversion to studies of behavioral genetics, he held off with publishing his famous book *Blueprint: How DNA Makes Who We Are* for 30 years (Plomin 2018).

For Robert Plomin, individual traits are largely a product of people's DNA. He does not shy from discussing his own genetic predispositions as discovered from analysis of his genome. His polygenic score[1] profile for psychological traits, which is the world's first such profile, is detailed in *Blueprint*. Many people panic at the prospect of using polygenic score profiles similar to those described in the preceding passage. Plomin, however, sees more opportunities than threats. In his opinion, it will primarily facilitate preventative measures for many diseases, education better adapted to the potential of each student, etc. He views himself as a cheerleader for these changes and an incurable optimist.

Professor Plomin, in the prologue of your book *Blueprint: How DNA Makes Us Who We Are*, you wrote that you waited 30 years to publish it because you were counting on the volume of research results collected during the time would make it easier for readers to accept the theses in it. Nevertheless, I have the impression that when the book was published you found yourself in the middle

[1]A polygenic score, also called a polygenic risk score, genetic risk score, or genome-wide score, is a number based on variation in multiple genetic loci and their associated weights. It serves as the best prediction for the trait that can be made when taking into account variation in multiple genetic variants.

of a firestorm of criticism at the point where ideology and politics mix with science (Comfort 2018). How do you feel in such circumstances?

I would disagree. I was, of course, very worried about how the book would be received. But I was quite pleasantly surprised by the response in contrast to the response I would have gotten even five years ago. Most of the reviews in the major media were very positive, whether in television interviews or in video blogs, as well as at around a dozen public events. I've been amazed at the positive reception. The only negative review I know of is one in *Nature*, but it hardly counts because it was by a historian rather than a scientist. He didn't even address the book, he just didn't like the message, and that review was so bad I think it actually helped because a lot of people were sympathetic and wondered how *Nature* could publish such a shabby piece of work.

Some of the critics claim that your theses represent the threat of a return to social Darwinism and eugenics or scientific racism. Could you respond to this criticism?

When talking to the public, I don't see that at all. Not one person out of hundreds has ever raised those issues. It's the media, and I think it's laziness. *Nature* published something around a year ago saying that modern genetics shouldn't be tainted with the history of eugenics because modern genetics in fact provides a much more subtle interpretation of genetics. Either way, I simply don't worry about it. I don't think people are concerned about a return to eugenics or determinism. If you understand the genetics, you won't think that it's leading to fatalism or determinism. These are issues I discuss in the afterword to the paperback edition of *Blueprint* that was published in June 2019. It presents my response to the reactions to the book and addresses these sorts of issues. As for eugenics, it bothers me a bit that people bring that up. As David Aaronovitch (2018) said in *The Times*, in what I think is one of the better reviews, it's interesting how quickly critics from a certain political persuasion jump with alacrity from the science to talking about ethical issues because they're not attacking the science. I don't think they can. They're just saying they don't like the message. It's a bit like the review in *Nature* where the reviewer says he doesn't like the idea of genes being important. So they're attacking on ethical rather than scientific grounds, which

I think is an anti-scientific stance. I don't necessarily have any trouble with that. What bothers me is that if you take it at its face, the Nazis were a totalitarian regime that killed lots of people with genetics as a rationale, but it was merely a fig leaf. There was no scientific justification for what they did, and most totalitarian regimes are environmentalistic. The science doesn't conclude that environmentalism necessarily excludes a murderous totalitarian regime—think of Stalin's Russia, Mao's China, or North Korea today. Those are totalitarian regimes driven by an environmental hypothesis that people are all the same and the state makes people into what they want them to be. So, I think the basic message is that totalitarian regimes are bad for people, and the scientific justification they use is just a fig leaf with no basis in reality.

This is a good point because the results of your studies also call into question the work of many psychologists who have spent their entire lives studying the impact of environment on human behavior. How do these researchers approach you?

I think that if I had written *Blueprint* 30 years ago when I was first asked to, I would have been crucified. The reaction was so hostile and there was this view that genetics is bad, while environment is good. However, in the subsequent 30 years a mountain of data has emerged that has convinced most scientists of the importance of genetics. This isn't just twin studies or adoption studies, but increasingly studies of DNA itself in unrelated people, and it's very hard to argue with DNA. So I think there is much greater acceptance of a more balanced view that recognizes the importance of genetics as well as the environment. Furthermore, I would strenuously object to the view that "environment is good, genetics is bad." I don't think genetics could ever do as much harm as environmentalism has done. Think back to 30–40 years ago, when the textbooks said that schizophrenia was caused by what your mother did to you in the first few years of life: toilet training, breastfeeding, that sort of thing. That mother blaming is incredibly bad because—imagine that your child is turning 20, they become schizophrenic and then you're told it's because of what you did in the first few years of life. Most environmentalistic theories involve parent blaming, and this is wrong. There's no evidence that early parental treatments cause schizophrenia. In contrast, there's strong evidence that genetics is one of the major reasons why some people become schizophrenic and others don't. I think the zeitgeist has changed and people are just more accepting of genetics. You hear it when people say "it's in your DNA, it's in your genes." There's no longer this knee-jerk reaction that says "genetics is bad, environment is good."

Would you say that this very popular conviction, especially prevalent among psychotherapists, that early childhood is crucial for who we are when we become adults is a myth?

I think that used to represent psychotherapy 20–30 years ago, but clinical psychology has changed quite a bit. It's become much more biosocial, recognizing biological and genetic influences. At first clinical psychologists were resistant for the same reason that educational practitioners are the biggest obstacle now, because they largely don't believe in genetics. Clinical psychology mistakenly thought that if things have a genetic component that puts clinical psychologists out of business, because if it's genetic then you can't do anything about it. But if anything, that's totally the wrong view about genetics—to say something is genetic doesn't mean you can't do anything about it; in contrast, it might actually mean you can do more about it. If you can identify particular genetic risks, perhaps you can work with that to improve treatment perspectives. This is because they realize it's not genetics versus environment, that there's a genetic component, there's genetic risk, but you need to study treatments that take into account genetic differences so that you look for treatment by genetic interactions. I think that's a hugely important and upcoming area.

Your research has also delivered results that demonstrate the uselessness of diagnostic manuals presently in use, such as the DSM. You yourself wrote in *Blueprint* that we cannot cure a disorder because there is no disorder, and so we can essentially toss our manuals into the shredder. Are clinical psychologists and psychotherapists taking advantage of the achievements of behavioral genetics and updating their methods of diagnosis and treatment, or are they still blindly groping around within the framework of the old model?

I see a lot of signs that clinical psychology, especially psychiatry, is much more advanced on this than some clinical psychotherapists who still believe in Freud and psychoanalysis, for which I see no evidence. I know there's a resurgence of interest in psychoanalysis, which I understand at some level of just understanding yourself, but as a treatment I don't see any evidence in support of it. In terms of genetics, I think it has huge implications for clinical psychology. The data very strongly suggest that, from the DNA point of view, we have polygenic predictors of schizophrenia and bipolar manic depression, and we will soon have DNA predictors of alcoholism, hyperactivity and all sorts of disorders. These polygenic scores involve thousands of tiny DNA differences into a multiple-gene aggregate. They're normally distributed across a perfectly normal bell-shaped curve. There's no evidence for any etiological break. So with schizophrenia, for example, there are thousands of tiny DNA differences responsible for the condition's heritability. That means we all have thousands of genetic risk factors for schizophrenia. There's no point at which you become schizophrenic or not. That probably has to do with the environmental stresses you're under, and so when you take a view of the whole it kind of does tear up our diagnostic approach, the medical model that begins with diagnosis that says "do you really have schizophrenia or not?" But the problem is that a qualitative dichotomous approach is just simply wrong. It's all quantitative. It's not a matter of "either or," it's a matter of "more or less,"—all quantitative, not qualitative. I think that's a very important concept to grasp. This is not to say that there are some people who have more problems than others. For example, in my area of educational problems: take reading disability, for example, which people try to dress up as a medical disorder by giving it a Latin or Greek name. You can call it "dyslexia," which makes it sound like it's more of a real disorder, and then the parent can say "doctor, does my child have dyslexia or not?" And it's just all nonsense from my point of view. It's all quantitative. Yes, kids have reading problems and we should work to deal with those problems, but let's not pretend that there is this discrete disorder that we then have to cure, and partly those cures involve neuroscientists looking for a hole in the brain, looking for the one thing that's gone wrong and made them reading disabled. Instead, it's almost a whole new perspective to say it's all quantitative. Some kids have reading problems more than others, and there's probably a genetic component behind it, but it's not a genetic component to have a reading disorder or not. It's all quantitative. And that's what I meant by "you can't cure it," because there's no disorder to cure. What you can do is alleviate the problem quantitatively. If a kid has reading problems, you help them to have fewer reading problems, you hope you'll help them to read better, but it's not like you're curing a disorder. You're dealing with a behavioral problem that you're trying to alleviate. It's really an anti-medical model approach. I have no doubt this is true, and it is already beginning to take over—remember how DSM-5 tried to shadow these diagnostic disorders, this nosology of classification, with a quantitative approach by saying "here's how you can measure these symptoms quantitatively." But there was so much vested interest in believing in the medical model of disorders that they scuppered attempts to include a quantitative approach. This was because if they had done that, and people collected quantitative data as well as these ridiculous diagnostic criteria of "you must have two of these symptoms for three months, and then you must have another symptom for six months," you could then ask empirically which approach works better in terms of prognosis and in terms of treatment: a quantitative approach or a diagnostic qualitative approach. So I think it's very exciting how genetics is already transforming clinical psychology and will continue to transform other areas.

How should contemporary diagnostic manuals look in the light of the newest research results on behavioral genetics?

There's such strong vested interest in these diagnostic manuals that there is very strong resistance against it, but I have no doubt in saying that in a decade's time those manuals will be thrown out. People will realize they're just pseudoscience. This is a difficult message because I'm not saying there aren't problems. There are people who have thought disorders, and there are kids who have reading problems. So I'm not dismissing those problems. My question is, "what's to be gained by pretending it's an etiologically distinct disorder?" I would say "nothing." In fact, there's a lot of harm done because then people think "oh, it's us normal people versus those schizophrenics," and it's not. We're all schizophrenic to some extent, it's just a question of how high our polygenic score is. This polygenic score of course isn't everything, but that's what's driving schizophrenia genetically. However, genetics only drives about half of the problem. So it's not to say it's all genetic, but I think it's a vital distinction that there are no disorders, there are just dimensions.

In *Blueprint* you wrote about generalist genes. What about the number of disorders? Do we need almost 400 as it is listed in DSM-5 now?

I think that's another important example of the way in which the new genetics is transforming clinical psychology. What we've found, and what we're discussing now, is that genetic effects are general across the dimension. They're not specific to a disorder. It isn't as though you have the genes for schizophrenia or not. It's all quantitative. In a similar vein, there's been another equally dramatic, almost unbelievable finding, that the genes for one disorder are largely the same genes for another disorder. The most striking example of this was with the first division you make when you're diagnosing a patient with psychosis. Up until DSM-5, you could either decide that a patient is schizophrenic or that the patient has an affective disorder like bipolar or major depressive disorder. Until DSM-5, you couldn't be both bipolar and schizophrenic, because that's the major first division in the nosology of psychopathology. And you might ask "well, who says?" The answer is, it's just people sitting in a committee deciding about these things. It wasn't really empirically based. And so the shock came when the first genes were found for schizophrenia—they were the same genes that affected bipolar manic depression. In subsequent work it's been found most genetic effects are general across all psychopathology. There are some genetic effects that are unique, but largely genetic effects concern one's propensity to have mental health problems. But the specific route that propensity takes, whether it's toward schizophrenia or bipolar, is probably mostly not genetic. I know that's a lot to take on board, but one of the hottest areas of research now involves *p*, which is a little *p* that represents general psychopathology; just like little *g* represents general cognitive ability. The idea of generalist genes first came about in research on cognitive abilities where we found that the same genes that affect spatial ability also affect verbal ability as well as other cognitive abilities. Genetic effects are general, and the specificity comes from the environment. So the hottest topic in clinical psychology and psychiatry right now is general psychopathology *p*, which captures most genetic influences across all types of psychopathology. People are even talking about trans-diagnostic treatments. That is, instead of thinking that we have to find one specific treatment for schizophrenia and one specific treatment for bipolar, researchers are thinking "if the genetic effects are general, maybe we need treatments that work across many disorders," because the problem isn't the specifics of schizophrenia or the specifics of bipolar. The genetic problem is what they have in common, which is a lot to get your head around. But this is another example of the way in which genetics is transforming clinical psychology. Again, it's going to tear up the diagnostic manuals that assume schizophrenia and bipolar are etiologically distinct disorders, because they're not. Genetically, they're mostly the same thing.

What you're talking about will certainly impact the ways in which therapists have worked until now. When discussing child-rearing, you often stress that what we can influence is behavior, but not inherited tendencies (an aggressive child will not cease being aggressive but can control aggressive behavior). Should we take a similar approach to psychotherapy? Should we abandon the widespread faith in cognitive psychology that our thoughts and beliefs influence our behavior?

Cognitive behavioral therapy (CBT) and a lot of other modern therapies aren't overly concerned with what caused the problem initially. They're more concerned about developing healthy cognitive habits for dealing with problems that arise, so instead of ruminating, it's basically about healthy or positive thinking. So I think that even without genetics, the forms of psychotherapy that work—not psychoanalysis, but therapies like cognitive behavioral therapies—seem more concerned about you simply dealing with your behavior. It could be your mental behavior as well, but it's not trying to change you fundamentally. It's trying to say "when you get anxious or your heart speeds up or whatever, and you start having panic attacks because you interpret those physiological feelings in a way that makes you think you're about to die, you don't need to think that way." I think this is happening already without genetics, but I think genetics will help it along. It's very important to emphasize that a trait or a dimension like panic attacks, anxiety, depression, even schizophrenia can show a strong genetic risk factor. That is, some people are more likely genetically to exhibit these problems than other people, but that doesn't mean there's nothing you can do about it. Causes and cures are not necessarily related, and I think that's what clinical psychologists believe, at least many that I know at the Institute of Psychiatry. They believe there can be a strong genetic component, but that doesn't mean you can't do anything about it. And conversely your treatments can work even if you don't know anything about the causes. For example, I have a strong genetic component tendency toward obesity, Body Mass Index. It doesn't mean I can't do anything about it. We all know we can lose weight if we just don't eat as much and exercise more, but knowing my genetic propensity actually helps me in my battle of the bulge to realize it's a lifelong fight to not put on weight and to try harder to lose weight. The main point here is that a connection between causes and cures isn't necessary.

Are we witnessing the DNA revolution in clinical psychology and psychotherapy?

Yes, I think it's already happening. As I say, a lot of the big studies, like the big randomized control trials of depression and CBT versus drug trials, they're all looking at DNA. If you have a large study and you're not getting DNA, I think this is a big mistake because the DNA is relatively cheap compared to everything else you do. For around seventy-five euros you can gather DNA, genotype it and bring a genetic perspective to bear on whatever it is you are studying. I know of a huge study in the USA that is looking at people with depression and how they respond to either CBT or drugs. There's been quite a bit of work on that already, but by collecting DNA they can now ask whether some people respond better genetically to drugs or to CBT. About hyperactivity, there are studies asking whether DNA can predict which kids will respond better to methylphenidate, and eventually you might be able to predict which children will suffer if they're given methylphenidate, which is the other side of the coin. Some children might respond very badly to drugs, and if you could predict who those people are, you wouldn't administer drugs to them. Once some of these findings come along, I think everybody is going to demand that they get their DNA genotyped. That's why there's a major discussion in the UK now about making genotyping available on the National Health Service.

We can improve health care in so many ways. That sounds optimistic.

I don't think it is optimistic, because as the Health Secretary of the National Health Service says, all of medicine is moving toward prevention, and to be able to prevent problems like severe heart attacks, alcoholism or obesity, you need to predict. DNA is by far the best early warning system that we have because you can predict from birth, which nothing else does. So I don't think it's optimistic at all. One severe heart attack costs the National Health Service about €750,000. You can predict that some people are at much greater genetic risk. The argument then becomes that it's actually unethical *not* to do this.

But some people are also afraid to know about their future.

Indeed, and there are some people who smoke two packs of cigarettes a day too.

Let's go back for a while to how the results of genetic testing are adopted by the scientific community. The resistance you experienced for years regarding acceptance of your research results, is perhaps a classic example of how the scientific establishment defends the official paradigms. It's something Thomas Kuhn called a "paradigmatic war," and which he considered a barrier to the development of science. Looking through the lens of his philosophy, would you say that psychology has undergone a revolution and a new paradigm has replaced the old one?

I think it's more a matter of science winning out in the end. There were some fields, say, 30–40 years ago, such as sociology, about which we could worry whether they would remain an empirical science. What I mean by this is that data are the ultimate arbiter—you have to accept the data. You may not like it at first, and it may be hard to accept a change to your old view that everything is environmental. That's a big shift to say "well, no, it isn't all environmental. It isn't just what your parents did to you. There's a strong genetic component." That's a lot for people to take on board, and that's why I avoided arguments with people. I just wanted to collect data, feeling that psychology will ultimately be an empirical science whose fundamental credo is that data rules. You have to look at the results of the data. You can argue against those results, but then more and more data arise leading to the same conclusion, which, as a scientist, you've got to accept. I think this is coming to a head with the replication crisis in science. Daniel Kahneman says that neuroscience is the poster child of failures to replicate. The results in behavioral genetics replicate. Nobody denies that. We have these huge findings about heritability being roughly 50% and about genetic influence increasing over one's lifespan, supporting the idea of a general genetic influence on environmental measures. These are massive findings, yet they replicate time and time again. You started asking about the resistance to genetics. I've written a paper on the replication crisis and why behavioral genetics results replicate (Plomin et al. 2016). Psychologists used to be so resistant to genetics. They made it hard to publish or to get grants about genetics. In a way, they were doing a service to the field of behavioral genetics because it made it much harder to convince people. As a result, you needed more and better data to do that. This explains why behavioral genetics results replicated, because you had to work very hard to convince people that these results were true.

Would you say that this crisis we are experiencing in psychology is also a product of insufficient attention being paid to genetic factors now researched?

I wouldn't say that in particular. I think it's the problems that people have been talking about forever: *p*-values, chasing probability values, and samples that are too small. Genetics only comes into play in terms of the etiology of individual differences. Much of psychology is what I call normative, asking about what the human species does; for example, asking when kids develop two-word sentences, a focus on averages. Most experimental psychology is also about averages; you take a group of people and you randomly assign them to an experimental group and a control group, then you do something to make them different on average. Individual differences in the ANOVA design are called error. What genetics is studying is that error. It's studying why people are different, and when you study individual differences you have to pay attention to effect sizes. Whereas if you study mean differences, there's a strong temptation just to look at *p*-values and not to ask about effect sizes. I think a fundamental reason for the replication crisis is the focus on averages and *p*-values rather than on effect sizes—how much variance is explained.

What should we do first to start overcoming this crisis?

We should start paying attention to individual differences and effect sizes. When you publish something and you find, for example, that boys and girls supposedly differ in verbal and mathematical ability. In these large samples you do get significant mean differences, but the effect size is less than 1%. So if all you know about children is whether they're a boy or a girl, you don't know anything about their verbal or maths ability. I think there's much to be said for focusing on individual differences rather than just treating them as a nuisance.

When discussing the crisis, there is frequent mention of the frauds that have been perpetrated by scientists in our field.

Indeed there is.

Considering your knowledge and experience, you can offer a credible assessment of a certain historical incident that remains a source of controversy today. Specifically, I'm referring to the case of Sir Cyril Burt. Considered a scientific fraudster, he arrived at very accurate conclusions with respect to inheritance of intelligence. Was he a master fraudster, a scholar too lazy to do research, or perhaps came up with his theories by chance? What do you think about his work?

I think that is a tremendous misrepresentation of the whole issue. That's old hat which has been discussed in three books. It's not at all clear it was fraud—that's a simplistic media trope to use. As we've shown in papers, you can forget Cyril Burt's data but still get the same results from the world's literature. This Cyril Burt story diverts from the science of what we're doing. Fraud contributes to the replication crisis, but I think it's a rather minor issue. My colleague, Stuart Ritchie, is writing a book on the replication crisis. Fraud captures a lot of attention, for example the current issues with embryos and in vitro fertilization. I think, however, that most of the replication crisis is at a more subtle level. It's not where someone just sets out to make up data, and for what it's worth, most scientists don't think that Cyril Burt made up his data. I don't think that's worth talking about anymore though. Let's forget about all his data and we still get all the same results. What I think is more important is how, for example, I hear scientists in large consortia where people combine their data saying things like "well, if we take the results for that measure, they look a lot better. Look, the *p*-value is less than 0.05. So why don't we just cut out this one sample and go with that other one? The results look a lot better." That's fraud. And that's where the failures to replicate come from, because once you start picking and choosing which data and which analytic methods to present, probability goes out the window. You shouldn't be allowed to talk about *p*-values anymore because you've destroyed their validity.

A good point. The growth in the importance of genetics in psychological studies has been accompanied by several other issues, such as epigenetics, which are becoming more popular among practitioners. With the help of epigenetics, outworn New Age concepts are being revived, some therapeutic practices are being justified, and attempts are even being made to revive Lamarckian ideas. Various contemporary gurus commonly employ the concept of transgenerational epigenetic inheritance to sell us on practices that involve the therapy of inherited traumas, or to work on changing our own genome. How much truth is there in any of this?

There's hardly any truth in it. People like anything that's anti-Mendelian and pro-Lamarck, and Lamarck was wrong. We don't inherit acquired characteristics. The notion of transgenerational inheritance is incredibly exaggerated. The few examples in mice (Dias & Ressler 2013) and in the Dutch famine study (Painter et al. 2008) do not prove that we inherit epigenetic marks. In fact, there are major processes involved in conception that get rid of any epigenetic marks from the mothers' chromosomal material. Gene expression is important, but it's a response to the environment. While I think epigenetics is valuable for understanding environmental influences from a biological perspective, it has nothing to do with inheritance. What we inherit is DNA sequence. RNA, which is what expression is involved with, has evolved to respond to the environment. DNA evolved to faithfully transmit evolved adaptations across generations. I find it amusing how people jump on the epigenetics bandwagon, but it's wrong to think epigenetics supplants genetics. We want to know everything that goes on in between genes and behavior, but this business about epigenetics somehow undercutting the importance of inherited DNA differences is nonsense from my point of view.

In discussing the genetics of individual differences it is impossible to avoid the highly controversial subject of gender studies. In the light of your research and your knowledge, would you say the claims

of gender studies about the cultural sources of sex differences are justified?

Gender is fundamentally biological in that females have two X chromosomes and males have a Y chromosome, but that doesn't tell you what the cause of the average differences between, say, boys and girls are. In the afterword of the paperback edition of *Blueprint*, one of the things I respond to is the criticism that I'm not talking about average differences between groups, like gender, class or ethnicity. I have good reasons why I'm not talking about it. It's very difficult to understand the causes of average differences between groups, whereas we have very powerful tools for understanding the causes of individual differences within groups. So I choose to study the tractable problem of individual differences. The other thing that's important to note here is that the causes of individual differences are not necessarily related to the causes of average differences between groups. For example, you can find that males are substantially taller in almost all cultures than females. This might lead you to think "well, that has to be genetic." But it doesn't have to be. In fact, the genes that are involved in height are not on the X or Y chromosome. They're on the autosomes, the 22 pairs of chromosomes that are not the X and Y chromosome, which are called the sex chromosomes. I think it's likely that the average difference between males and females in height is genetic, but conceptually it doesn't have to be. It could be that the genes for height or hormones work differently in a girl's brain than in a boy's, and that could be driven environmentally. Perhaps height is a bad example to try and come up with an environmental explanation. The larger issue is that in developmental psychopathology we find some of the biggest differences between males and females—think of autism, hyperactivity, or even reading disability. It seems like anything with a neurological basis is worse for boys than girls. While you might think it's genetic, every attempt to find a genetic basis has failed. It could well be something in the environment, or that the genes involved work differently in a boy brain and a girl brain, given that they differ, for example, in hormones. The issues around gender studies and some of the politics that are involved in it are interesting, but I refuse to talk about average differences between groups because, as I've said, I don't think we have very powerful tools for nailing down the causes of the differences. I think that's why there's so much heat and so little light. Also, as I say in the afterword of the paperback edition of the book, I don't have to study everything. I'm increasingly comfortable saying that if you want to talk about group differences, that's fine. I've got enough on my hands with individual differences and I take enough flak for studying individual differences that I don't need to stick my neck out and study average differences between groups.

Another hot topic is artificial intelligence. Is the growth of this field also opening up new research perspectives on behavioral genetics?

Not to my knowledge, because artificial intelligence is a normative approach, it's not talking about individual differences. It might eventually be able to do that, but right now artificial intelligence means a lot of different things. For some people it involves analytics, a sort of approach to problems, like using advanced analytic techniques such as mathematical modeling. And those analytic approaches are being used in genetics, for example, to try and find genes when you've got billions of DNA bases involved and you want examine interactions between genes. But if what you mean by artificial intelligence is robots, I don't see where that's going to help us in behavioral genetic research, although robots have greatly increased the speed and lowered the cost of genotyping. I hope behavioral genetics will be open to new approaches. If you can show me something that will help me try to find genes related to behavior, I would welcome it. For example, machine learning algorithms are increasingly being used to solve the very complex problem of trying to find all of the thousands of DNA differences and their interactions that affect behavior.

We have devoted a lot of time to analyzing controversial issues related to genetics and psychology. Let's talk about the achievements of the field. What great questions has psychology as a science managed to answer so far?

To be fair, I'm a one-trick pony. So I immediately think of genetics and how drastically psychology has been transformed. It's been a huge step to go from Freudian thinking that everything is environmental, that we are what we learn, to recognizing that inherited DNA differences account for more variance than anything else we know about. It accounts for something like half of the differences between people. This is a big advance, and then the DNA revolution came along and showed that we can identify some of those DNA differences and begin to predict behavior from DNA alone. And that is an earth-shattering development in psychology. It's what I love about behavioral genetics, something Peter Urbach and other philosophers of science talk about as a progressive science that builds on previous results rather than hopping from one fad to another. But if you look at recent issues of psychological journals, a lot of it doesn't seem very progressive. It seems like a lot of fads where something is of current interest and people rush that way, and then a few years later people get bored with it. It's not that they've solved the problem, but rather they just head off in another direction. I think a lot of psychology hasn't been progressive. Although it's self-serving and it's my area of interest, I don't think any other field in psychology has caused such a transformation over the last few decades as genetics has done.

Are there any great questions which remain unanswered?

From my very limited perspective, the big question everybody's facing is not how we find some of the genes that predict the heritability of behavior, but how we find all of them. That's going to involve new technology. Instead of dealing with SNP chips, which are DNA arrays that measure a few hundred thousand single nucleotide polymorphisms across the genome, we're now moving toward whole genome sequencing, which identifies the sequences of all three billion nucleotide bases of DNA; this captures everything that you've inherited. The rest of the DNA variation must be there, and along those line there have been some big developments in recent months. I think the next big thing is whole genome sequencing, which I believe will make it possible to predict most of the genetic variability in behavior.

What questions are you trying to find answers to these days?

Most of my work has shifted from what we call quantitative genetics, which involves twin and adoption studies, to exploring the extent to which inherited DNA differences are important. For the last 20 years, and especially in the last decade, I've been much more interested in using DNA itself to make these predictions. The neat thing about polygenic scores is you can use them to make genetic predictions for any sample of unrelated people, without needing twins and adoptees. That's going to make genetics much more accessible to psychology. And as I've said, if you have a very large and valuable sample and you're not collecting DNA, you're making a big mistake. You don't have to become a geneticist. You can simply use DNA to put a different perspective on whatever question you're asking. It's really just the beginning of genetics in psychology, and I strongly believe that genetics is going to transform not only clinical psychology but also society in general and how we understand ourselves. 25 million people have already paid about 150 euros to have their DNA genotyped because there's a real interest in this, and I think that interest is going to increase as we improve at predicting psychological traits from DNA alone.

Probably our conversation will also be read by students and young psychologists just getting their careers started and who are wondering what specialization to explore. What areas of investigation in psychology do you think are particularly promising and would recommend exploring?

Psychology is the most popular undergraduate major in the UK, and I think that's because the wide range of questions that we address in psychology are questions that are of interest to all people. And even though I'm in a medical school now, so many people in medicine are coming around to the idea that much of the medical burden on society has to do with psychology. People come to the GP and sometimes they have a very specific problem, but very often it has to do with depression and psychological sorts of issues. That's why I think psychology is going to remain an increasingly important field, not in itself as much as across all the life sciences. So I naturally think that psychology is important, but I would say that anyone starting out in psychology who doesn't keep an eye on genetics is making a mistake, and psychology departments that aren't teaching their psychology students about genetics are doing them a disservice. Genetics is going to become increasingly important and I would urge young people to learn about genetics even if they're not taught about it, because it's going to change all aspects of psychology. It doesn't matter whether you study perception or the nervous system. Whatever you study in psychology, especially clinical psychology, you'll make bigger advances by considering genetics.

At the beginning of your career you yourself choosing the subject of your research at the same time decided to swim upstream against the dominant tendencies of the time. Would you encourage young people to adopt a similar attitude?

People shouldn't just be contrarian and say "everybody's going this way so I'll go in the opposite direction." The reason I do what I do was that I was so impressed with the power of genetics that I couldn't understand why I'd never heard about genetics in my undergraduate psychology, or even in graduate psychology, up until the point I had a course in behavioral genetics. If you do find something that interests you, go with it, even if it isn't currently in favor in psychology. Particularly if you have the right disposition, by which I mean you have to be strong to go against the flow and be prepared for people not liking you. That's a small price to pay for being right.

Who would you point to as an example to follow?

I had a lot of heroes, but for me it wasn't about just following someone else. I always did my own thing, but there were people I admired for the way they wrote and for the cleverness of their ideas. My religion is science, not individual scientists. For example, when I write, you may notice that I never use anybody's name. I've never said "Jerry Kagan said this," as some typically do, talking about people who are famous because it gives credibility to what they're saying. I believe people belong in the parentheses, in the brackets at the end of the sentence. You can say something and you can cite a study, but you're focusing on the ideas rather than the person. That has always been important for me—I believe in science, but I try to resist the temptation to believe in scientists and to hold them up as heroes because I don't think the people are important. I don't think I'm important. I think my research is important, and in the end that's what counts. If you're too focused on people and personalities, I think you start worrying about awards and getting credit. I see this happening to a lot of people as they get older—they're concerned about not getting proper credit for what they think they discovered. Well, they're not important, what's important is the discovery. That makes us feel part of a community of science instead of a cult of personality of scientists. But that goes against a strong evolutionary trend—I think we do have this tendency to want to have heroes and to make it about people rather than about ideas.

That goes strongly against our individualistic culture. What should young people be careful of? Can you give them any other tips?

I'm not so good with giving advice. I like to get the brightest students I can, those who are hardworking and really interested in being scientists, and often we get the best students from all over Europe. One criterion I use in selecting students is that they're feisty—they don't agree with everything I say. They don't take my advice, and I think that's a good thing, to be a free thinker and to argue with people. But then if you're wrong, you have to be gracious and say "yes, I was wrong about that." I'm not really one for giving people advice except to say "don't get locked in too early with a very specific area." My general strategy is like in chess, to play the board broadly at first and keep as many doors open as possible. In psychology, that would involve embracing new techniques that come along, because the field is moving so fast. If you become too specialized too early, it's like Darwinian evolution. If that niche that you've very specifically evolved toward continues to be important then you might be a big fish in a little pond, but chances are the field will move along and you're left in an evolutionary backwater, specialized in something that's no longer of interest. I would encourage people to go at it broadly, which is why I like the American model of graduate school in psychology, where it's assumed you don't know anything. Even if you hold a psychology undergraduate degree, you're taught everything again at a graduate school level for the first two years of training. Whereas in the UK and I think in a lot of Europe the undergraduate psychology degree is viewed as the final degree, and then graduate school is viewed as a specialty. You're supposed to come in and say "what I want to do for my Ph.D. is this very specific project." I think that's a big mistake because you don't know enough after an undergraduate degree. What I like is for my graduate students to end up doing a Ph.D. on some topic they didn't even know existed when they got started. So, while I can say that I don't like giving advice, that's the advice I do tend to give my students, even though they often don't pay attention to my advice anyway.

Is there anything I haven't asked you about during this conversation that you would like to mention?

There's quite a lot to discuss regarding the new developments in DNA, but I think from a psychologist's perspective we've probably touched on the biggest topics. I suggest that readers of your book read the afterword to the paperback edition of *Blueprint*, because while we have touched on most of the issues that have come up since the hardback was published, there are a couple we haven't.

Would you agree to answer one of the 30 questions that my readers would like to ask the most eminent psychologists in the world?

Yes.

Please draw one.

I'll take number 12.

Interesting question: What do you believe is true, even though you cannot prove it?

I don't believe anything is true. If I can't prove it, I could say it's a hypothesis. And I do have hypotheses, such as we will succeed in finding all the DNA that accounts for the heritability of behavior. But that's just a hypothesis. I would say one of the phrases I most overuse is "it's empirical." You can say anything and hypothesize anything you want as long as it's empirical, because in the end it's the data that will determine whether it's true or not. You can't simply *believe* something's true.

Do you apply the same philosophy to your private life?

Yes. That's why I stopped being a believer. I became an atheist at the age of 12 for that reason. I don't believe in religion, because you can't prove it.

Selected Readings

Ashbury, K., & Plomin, R. (2013). *G is for genes*. New York: Wiley Blackwell.

DeFries, J. C., Plomin, R., & Fulker, D. W. (1994). *Nature and nurture during middle childhood*. Oxford: Blackwell.

Dunn, J., & Plomin, R. (1992). *Separate lives: Why siblings are so different*. New York: Basic Books.

Petrill, S. A., Plomin, R., DeFries, J. C., & Hewitt, J. K. (Eds.). (2003). *Nature, nurture, and the transition to early adolescence*. New York: Oxford University Press.

Plomin, R. (1986). *Development, genetics, and psychology*. Hillsdale, NJ: Lawrence Erlbaum Assoc.

Plomin, R. (1994). *Genetics and experience: The interplay between nature and nurture*. Thousand Oaks, CA: Sage.

Plomin, R. (2004). *Nature and nurture: An introduction to human behavioral genetics*. Boston: Wadsworth Publishing.

Plomin, R. (2018). *Blueprint: How DNA makes us who we are*. London: Allen Lane (2019 paperback; Penguin Press).

Plomin, R., DeFries, J. C., & Fulker, D. (2006). *Nature and nurture during infancy and early childhood*. Cambridge and New York: Cambridge University Press.

Plomin, R., DeFries, J. C., Knopik, V. S., & Neiderhiser, J. N. (2017). *Behavioral genetics* (7th ed.). New York: Worth Publishers.

Plomin, R., DeFries, J. C., McGuffin, P., & Craig, I. W. (Eds.). (2002). *Behavioral genetics in the postgenomic era*. Washington, DC: American Psychological Association.

Reiss, D., Neiderhiser, J. M., Hetherington, E. M., & Plomin, R. (2003). *The relationship code: Deciphering genetic and social influences on adolescent development (adolescent lives)*. Cambridge, MA: Harvard University Press.

References

Aaronovitch, D. (2018, September 29). Robert Plomin interview—Why genetic testing is the future. *The Times*. Retrieved from https://www.thetimes.co.uk/article/robert-plomin-interview-why-genetic-testing-is-the-future-m2gcskpkv.

Comfort, N. (2018). Genetic determinism rides again. *Nature, 561*, 461–463.

Dias, B. G., & Ressler, K. J. (2013). Parental olfactory experience influences behavior and neural structure in subsequent generations. *Nature Neuroscience, 17*(1), 89–96.

Gottfredson, L. S. (1994, December 13). Mainstream science on intelligence: An editorial. *Wall Street Journal*.

Herrnstein, R. J., & Murray, C. (1994). *The bell curve: Intelligence and class structure in American life*. New York: The Free Press.

Painter, R. C., Osmond, C., Gluckman, P., et al. (2008). Transgenerational effects of prenatal exposure to the Dutch famine on neonatal adiposity and health in later life. *An International Journal of Obstetrics and Gyneacology, 115*(10), 1243–1249.

Plomin, R. (2018). *Blueprint: How DNA makes us who we are*. London: Allen Lane.

Plomin, R., DeFries, J. C., Knopik, V. S., & Neiderhiser, J. M. (2016). Top 10 replicated findings from behavioral genetics. *Behavioral Genetics: Perspectives on Psychological Science, 11*(1), 3–23.

8

Susan J. Blackmore: Parapsychology, Memetics and Consciousness

What could be more interesting than trying to understand the mind that is trying to understand itself?

Susan J. Blackmore photo by Adam Hart-Davis

T. Witkowski, *Shaping Psychology*,
https://doi.org/10.1007/978-3-030-50003-0_8

Have you ever looked at some people and wondered how much electricity they could produce if you connected them to some sort of generator? The energy that radiates from Susan Blackmore would almost certainly light up a small town, and if anybody decided to put together a ranking of the most energetic people on the planet, she would doubtlessly make the list. She continually gifts this energy to the people surrounding her, and it seems fair to say she's got a tremendous surplus of it when considering that, alongside her exceptionally active professional life, she also plays in a samba band, and enjoys power lifting, painting, kayaking and gardening.

While as a teenager Blackmore had already begun posing questions about supernatural forces that science did not particularly understand, her interests in this area really took off when she pursued her studies at Oxford University. She describes one particular incident that she considers a breakthrough moment thus:

> Within a few weeks I had not only learned a lot about the occult and the paranormal, but I had an experience that was to have a lasting effect on me—an out-of-body experience (OBE). It happened while I was wide awake, sitting talking to friends. It lasted about three hours and included everything from a typical „astral projection," complete with silver cord and duplicate body, to free-floating flying, and finally to a mystical experience. It was clear to me that the doctrine of astral projection, with its astral bodies floating about on astral planes, was intellectually unsatisfactory. But to dismiss the experience as „just imagination" would be impossible without being dishonest about how it had felt at the time. It had felt quite real. Everything looked clear and vivid, and I was able to think and speak quite clearly. (Blackmore 1987)

This experience inspired her to engage in an intensive search for the essence of paranormal phenomena. After spending time in research on parapsychology and the paranormal, her attitude toward the field moved from belief to skepticism. This adventure in Susan's life lasted around 30 years.

However, the thing that brought the greatest recognition to her in the world of science was the popularization of Richard Dawkins' notion of the meme (Dawkins 1989), and the formulation of a theory of memetics she presented in *The Meme Machine* (Blackmore 1999), later elaborated in her conception of consciousness. She says about this:

> When I say that consciousness is an illusion I do not mean that consciousness does not exist. I mean that consciousness is not what it appears to be. If it seems to be a continuous stream of rich and detailed experiences, happening one after the other to a conscious person, this is the illusion.

From the memetic point of view, the self is a memeplex that does not serve us in making decisions, but exclusively to disseminate the memes that comprise it.

Susan Blackmore considers herself as one of those scientists who are incapable of separating their scientific views from life. She cannot understand the attitude of biologists who practice their science during the week and then head for church on the weekends, or physicists who believe they will enter heaven after death. This is also why she actively seeks means of protecting herself from the tyranny of memes. The defense she has practiced for decades is Zen meditation—the cleansing of thoughts and concentration on the "here and now," a life without the false feeling of self.

She is an unusual scientist, because at a certain moment in her career she abandoned a safe job at a university in order to strike out on her own. Some of her coworkers at the time considered this a moment of insanity, while others were sincerely jealous. She never returned to academia, yet despite that fact she consistently features in rankings of the most outstanding living psychologists.

Professor Blackmore, among well-known scholar-psychologists, you are probably the only one who quit working at a university in order to follow your own freelance path. Was this decision a manifestation of dissent directed against the norms and relations prevailing in academia?

If you mean back in the 1970s, then I never thought of myself as having a career at all. I didn't look to the future, but rather I only wanted to pursue my obsessions with trying to understand the mind. Initially I just wanted to prove to the world that psychic powers such as telepathy and clairvoyance existed. When I found out that it almost certainly does not, I moved on to a deep bewilderment about the nature of consciousness. I then worked on my own for many years, raising my children, and doing writing and research when I could. Later, after ten years of an actual university job, I gave it up to return to being freelance. This time, when I gave up my job, it was to some extent about academia and the increasing workload, the growing number of less-able students, and the pointlessness of red tape and meetings. I felt like I was being paid not to work! I prefer the freedom to work on my own as much as I want, and I'm happy to cope with less pay and more uncertainty in exchange.

I can imagine that choosing this difficult and perhaps somewhat romantic career path could lead to some negative consequences. Which of them have you found to be the most burdensome?

First of all, I never thought of it as a career, so in a way I'm rejecting the premise of your question. But did my choices have negative consequences? Well, yes. For a lot of my earlier life, when my kids were small, I didn't have a job. I didn't expect myself to be able to get a job because of the work I was doing on telepathy, clairvoyance and out-of-body experiences. I got tiny little research grants, and my husband had a lectureship, which I was sort of jealous of, but sort of not. We had very little money, and I was at home bringing up the kids most of the time, trying to write my books in whatever little time I had with two young kids. I suppose that was the most negative aspect of it. Otherwise, I don't find anything negative in it. Of course, I got a lot of flak from believers. Interestingly, when I was a believer I never experienced any nastiness from the skeptics, but when I became a skeptic I got lots of nastiness from the believers. It's the believers in the paranormal and life after death who are really vicious in the hate mail and everything. Sad, isn't it?

Yes, that's true. But your position in the scientific community is very strong, your name appears in rankings of the most outstanding psychologists. What do you think—would someone less recognizable be able to enjoy a similar career as an independent thinker in the modern world? Is it conducive for people who want to work independently?

I don't know. I think it has always been difficult, and it still is now. One thing you might find odd is that it's easier for a woman to do it if you are the kind of woman who wants to have children and stay at home and look after those children. You're forced to find a way to combine your work with your home life, and I was always very happy to do that. Also, when the kids were young, I was living in a village where most of the women were at home all the time. They weren't writing books, but it was quite normal for women to be supported by their husbands. Nowadays it's much easier for women to get a job than it was in my time, but it's kind of harder in a way, because that's expected of them and it's not so easy to limp by on very little money and use your husband's income. That said, I don't know why my name comes up in those rankings, because I haven't done any fantastic psychological work. I've done quite a lot on consciousness and written a textbook, but I don't have any great theory of consciousness that solves any problems. All my work on the paranormal ended up showing that there were no phenomena and providing alternative explanations. I think one of the reasons I became well known is that in my parapsychology days, radio and TV were all there was. There was no social media or internet, and maybe I just had the kind of voice, or presence or enthusiasm—whatever it is—that they needed. There were always programs looking for a skeptic for "balance." So I went on programs with a hundred people who'd seen a ghost and Sue Blackmore to say it was all in their mind, or 500 people in the studio who had been to heaven and come back and Sue Blackmore to say it's all brain mechanisms and hallucinations. I did that, what I call "rent-a-skeptic" for many years, and while I earned a little bit of money that way, I certainly earned a lot of recognition. If you think of the important psychologists on those lists, you can probably say for nearly all of them: „yeah, that's the person who discovered X, or this person created a transformative theory," but I don't think you can really say that about me.

You're simply being modest.

I try to understand because I think it's odd.

Do young people have opportunities to fulfill their ambitions in a similar manner?

Well, there are always opportunities. If you become obsessed by something, then you'll do it, whether it's today or decades ago like me. You will find a way because you care so much—you might be a philosopher obsessed with understanding the mind-body problem or whatever it might be, and you want to know. Or you might be a biologist so disturbed by climate change and mass extinction that nothing else matters in your life. If nothing else matters in your life, you will find a way, and there will always be people who do that. Of course, now in order to be heard you need to be out there on the internet, which takes a lot of time and effort. I don't do Twitter, Instagram or anything like that, only a rather simple Facebook page, and my own website that I started back in the 90s and still maintain now. It's a vast thing and I'm happy to do that. Nowadays you have to go in really hard if you want your work to be seen. But again, if you're really obsessional about something, you just get on with it. If you're lucky and your ideas are good enough then you'll flourish, and if they're not you won't. I don't think that's changed even though the world has changed.

Should contemporary societies provide conditions for young, individualistic intellectuals different than those at today's universities?

I think they are there if you want to find them. My knowledge is more about science than the humanities and other subjects, but you're not going to get anywhere with any of this thinking if you don't go to university in the first place. You must have some basic training in whatever your subject is, whether it's theoretical physics or biology or whatever else it might be—you have to start there, and then there are lots of ways to branch out. I don't know what it would mean for a society to provide other ways. I'm trying to think of a better word than the "crazy people," but people like me just have an obsession we want to go off and understand. We'll find a way in any society. I don't know what you could provide. It's not like a business where you can provide startup money or places to work or something like that. If your subject requires a lab, you've got to be somewhere that has a lab, whether that's a university or a private institution or a corporation. You need the equipment and you need the money to do that. If you're a philosopher, or a psychologist like myself not in the lab, just thinking and reading and so on, then you just need a house to live in and you can get on with it yourself. I'm not sure how to answer that question. What sort of things do you think society could provide for such people?

I don't mean any specific expectations. I just wonder what you think about it.

Well, it's a very interesting and difficult question. But I hope we are still living in a world in which good ideas flourish. And I get really depressed because such a load of absolute crap flourishes, and we have Donald Trump coming to England today. Oh, God… And we have so much fake news, and we have Gwyneth Paltrow and her wretched false claims about health, which ought to be illegal and yet she gets away with it. I can get depressed about that. Nevertheless, if you think deeply about problems in math, physics, biology or psychology and come up with really good ideas, you'll find a way. Those ideas will get out, they'll escape. That's my hope, that it was always true and remains true today.

We began this conversation somewhat unusually, for the simple reason that, in my view, your individual career path and choices serve to highlight the problems that many contemporary thinkers are faced with. Now I'd like to discuss the things which you have taken up in your life and work. As you mentioned before, you devoted 30 years to researching paranormal energy, only to abandon the field forever. What led you to walk away from it?

I tried to walk away a long time before I finally managed it. I began pursuing my Ph.D. in 1975, and I'd already done some experiments before that. So I began serious experiments then. By 1980, I had come to the conclusion that there were no paranormal phenomena. I wouldn't say a hundred percent, though. They may exist and I just couldn't find them, but I'm as sure as I can be that there are no such paranormal phenomena. So I'd already become very skeptical by then. I kept going because the claims kept coming, and because I knew a lot about the experimental methodology and the statistical methods used in that field. I felt I could contribute, and then people kept asking me what I thought about all sorts of experiments, so I'd get lured back into it. I tried to walk away several times, and it really wasn't until the year 2000 when I simultaneously left my job and the field of parapsychology in order to write my big textbook on consciousness. I just wanted to get away. Many people would have loved my full-time job as Reader at the University of the West of England, but I gradually cut down to two days a week and then left altogether. I have always been happiest being at home, being able to work all the time with no one to deal with at all, just reading, working and writing alone, and I had I had the determination to write. Around 1999, the Internet appeared and I started my website. I also used to write for *The Guardian* and *The Independent*. So I wrote that I was leaving, and one article "Why I'm leaving" became very popular. At the same time I finally escaped from parapsychology. Since then, if people ask me questions about particular psi experiments, I say that I can give an informed opinion on anything from way back in the 80s, but about recent stuff I cannot. This is because the process of trying to explain someone else's results and find out what went wrong in some experiment—if anything did go wrong—is often harder work than doing the experiments in the first place. It's also terribly depressing. And I just couldn't bear it anymore. So it took me about 20 years to escape.

Considering your familiarity with researchers studying paranormal phenomena, would you say that many have arrived at conclusions similar to yours—that such phenomena do not exist—but differently from yourself, they aren't strong enough to walk away from a lifetime's worth of work?

It's an interesting question. There are lots of different things going on there. First of all, when I was involved in parapsychology I knew many good, bright researchers who came to that conclusion and just left. They might have been in the field for a few years, done some research, came to the same conclusion I did, and went off and did something else. Julie Milton was very good researcher and she has done some more parapsychology, but she went off and got a career elsewhere. And Debbie Weiner was another American who left to do other things. There are others who couldn't make up their mind, but on the whole the human mind does tend to go one way or the other on big issues like this and I would say the vast majority of people in parapsychology know a lot of the criticisms. They don't accept them. They have such a strong belief in paranormal phenomena they believe to have experienced themselves that they are simply able to dismiss criticisms from people like me. There are a very few who are so determined to be right about the existence of paranormal phenomena that they will actually cheat in their experiments. Not many, but unfortunately even a very few can cause enormous damage to a field and that, sadly, has happened.

I've read and heard on several occasions that despite abandoning parapsychology, you continue to believe that we should support research on paranormal phenomena. For many skeptics this is a very surprising attitude. Could you elaborate on your views?

Yes, if you discover that some theory in physics or chemistry is wrong, then it just gets forgotten. You don't have to suppress it, people won't give money to it, and it goes away. Claims of the paranormal are somewhat different because so many people have experiences that they are absolutely convinced were psychic. It can be the simplest thing that happens to all of us—you think of a friend, and 10 minutes later they ring you up, and you think "oh isn't that strange?" Or you wake up after a really vivid dream of somebody, and you discover that they died the day before and you didn't know about it. The statistics have been done and that's bound to happen to quite a lot of people just by chance. Without going deep into the weeds, people have these experiences, which we know from some of the surveys I did a long time ago. We know that the major driver of people's belief in the paranormal is their own experiences, and it doesn't help to say "well, that's just a coincidence. You know, that's probability." We all are very bad at understanding probability, and it doesn't wash with people to say that. So it behooves us to provide some kind of explanation for people's experiences. That's why I have gone away from testing paranormal phenomena to looking into why people have out-of-body experiences and other kinds of odd things that happen to them, because that can help them abandon belief in the paranormal when they understand how and why tunnel experiences and ghosts under the bed occur. All these things which we now understand very well can help a lot of people in letting go of those false explanations. But it is still conceivable that there is some kind of psychic connection somewhere, and one could wonder: have we got it wrong about the nature of the universe? It's a vanishingly small possibility, but it's nevertheless a possibility. It would be so important to science if there were psychic phenomena, so I am very glad that there are just a couple of labs here and there which do get money and carry on investigating, because if we didn't look we'd never find it. I don't think we'll ever find it. I don't think it's there, but I'm really glad that there are at least a few people going on looking, because it would transform everything in science if it were true.

That's absolutely true. What do you think—apart from parapsychology, are there other fields within psychology where the consistency trap and engagement keep sway over other scientists who, despite the absence of results, continue to pursue barren lines of inquiry?

I don't know of anywhere exactly like that. There is a big crisis in psychology at the moment because it's very easy—particularly in the modern world with very fast communications—for a psychologist, if they are well-known, to make some claim and then everyone just believes it. It's very, very hard then to get that out of the public consciousness. A very simple example is a famous experiment done which shows that if you talk to a group of people about old people and things to do with old people and so on, and talk to another group of people about youth and young people, individuals in the experimental group would walk down the corridor more slowly if they were thinking about old age than if they were thinking about youth. This is such a lovely idea that everyone believed it, but then when people tried to replicate it, they couldn't. I don't know what the latest is on this. But, of course, in that example and many others, it's not that people try to keep on doing the same thing when they don't get results—they'll do enough of that and then change over and do something else. People will, however, go on bashing their head with the same theory, but that's a very different argument. People become wedded to their favorite theory and they will go on and on, and there's that famous statement that in order to have a new theory you just have to wait till the people who believe the old one die. Well, things are going too fast for that now, but it's still true to some extent. People will cling to their theories; this is absolutely true in parapsychology and it's true in a lot of psychology as well. It's psychology itself, it is hard work to change your mind, and I was forced to do it very young. I was around 25 when I was really confronting the possibility that all my work up till then had been a waste of time. It wasn't really a waste of time, but all I had discovered was that I couldn't discover any paranormal phenomena, and when you're 25 and you've done five years of work that seems like an awfully long time. I was forced to change my mind and was young enough to do it, so I'm not afraid of changing my mind. That said, it's painful rather than pleasurable to think "I was wrong. I spent all this time doing these things which got me nowhere. I stupidly believed this and proclaimed this, what an idiot I am," and find another way, so I understand quite well why it's so hard. Nevertheless, the heart of being a good scientist is changing your mind when you're wrong, and caring more about evidence than about whether your own theory is better than somebody else's.

Yes, it's very difficult. Let's now move on to another topic, that of memetics. What were the circumstances in which you decided to pursue memetics?

That's a much easier question than the previous ones. What happened was, in 1995 I was experiencing chronic fatigue for the first time. I'd had an exhausting summer and autumn, I fell ill in October, and Dan Dennett's book *Darwin's Dangerous Idea* had just come out (Dennett 1995). I had to spend an awful lot of time in bed, and I began reading that book. I had read *The Selfish Gene*, in which the term "meme" was originally coined, back in 1976, but I'd forgotten all about it (Dawkins 1989). Dennett's book reminded me about memes. At the same time, my Ph.D. student was filling in for me during some of my lectures at the university. I saw quite a lot of him, and he wrote an essay for me about memes and consciousness. Those two things combined had me there lying in bed thinking about memes. Because I was a very slow reader due to my illness, I could only read a little bit at a time. I was lying in bed most of the time for six months or so, staring at the ceiling a lot because I was too tired to read or get up, I couldn't type or do anything like that and there were no mobile phones in those days. I was cut off from the world, and I just thought and thought and thought, and all these ideas came to me. It went like this: if you think of culture as a second replicator, if you think genes are the first replicator propagating for their own benefit, trying to get copied whenever they can, and you then transfer that Darwinian idea to culture and say "gosh, every idea, habit, skill, story, song, whatever that I know, is information competing to get into my head to stay there and get propagated into somebody else's head," suddenly so many things in the world began to make sense. It was a revelation. The same revelation that often people get when they understand Darwin in the first place, when you suddenly grasp natural selection and how evolution works and you realize that design comes out of nowhere by this process of copying, varying and selecting. It was a rather similar revelation to me about memes—suddenly the world looked different, and it's always looked this new different way to me ever since. So I had to bottle up these ideas because I couldn't write anything. I just turned over ideas endlessly in my head. Notions about the origins of language and how language was a parasite in the first place, not like all sensible biologists who think that language evolved because it was adaptive for genes. Or about why people are altruistic; or why we have such huge brains because memes drive us to that, this was why I talked about memetic drive. All of these ideas were swarming around in my head and it wasn't until the following summer, about a year after I first got ill, that I was able to begin writing. That was when I wrote *The Meme Machine*. So it was actually an illness that enabled the thinking that went into those ideas.

After the publication of *The Meme Machine*, memetics began enjoying massive popularity among scholars across various fields. Was this also reflected in the volume of empirical research?

No. There's a big mismatch here. Memetics is absolutely not popular at all within the fields which it ought to apply to. There are probably less than half a dozen biologists that I know of who take memetics seriously. There are several very good books on cultural evolution at the moment. The particular one I'm reading now by Joseph Henrich (2016), *The Secret of Our Success*, claims that it's because of culture we're so different from every other species. Well, that's what I said in *The Meme Machine*, but he does not take the view that memes are a replicator, and most other biologists agree with him. The word "meme" is not even in the index of that book, nor is my book cited there. So, within conventional biology and anthropology, cultural evolution has become a respectable field of study but memetics has not. The memetic slant on it is to say that all these cultural items which we call memes are a replicator competing to get themselves copied, and the consequences for us and our culture are a product of that competition. Whereas the standard view is that cultural evolution is ultimately in the service of genes and we don't need to think in terms of a second replicator. So that's all going on within biology. Memetics is not understood at all by the public because most people think that an Internet meme is the only sort of meme there is. We have a lot of problems here.

That's true that we misunderstood the notion of memes because of Internet memes, but I remember that memetics was quite popular, and that much hope was placed in memetics as a conception unifying the achievements of the scientific study of people and of culture. In your view, has it lived up to those expectations, or will this perhaps occur in the near future?

No, it hasn't lived up to these expectations, partly because it's very, very difficult to make empirical tests to distinguish between memetic theory and standard cultural evolution theory. I am very disappointed and sad that I have not been clever or insightful enough to come up with such experiments. But also, at my age, I'm no longer interested in doing lab work or fieldwork myself, and there are so few people in the world who take a serious view of memetics that they're not doing it either. I don't think it's impossible. I also like to think about the fact that, going back to Darwin in the first place, nobody could really make sense of his theory of evolution by natural selection until much later, when we began to get a grip on genetics, and that memetics may need a breakthrough like that. But recently I've been thinking it's time to start writing about memes again, and I reread *The Meme Machine*. Richard Dawkins reread it too and wrote some wonderful tweets about how important a book it is. What struck me when I was reading it was that an awful lot of the things I said in there have come true in some way, and you can't easily say "oh, well this proves it" because there are plenty of other theories around. But it has encouraged me a lot, and I think the whole advent of Internet memes has demonstrated the power of it. To me, what's important about memetics is the difference between memetics and other theories, that memes are competitive; information is competing to get copied, and memes are the final arbiter. That's what it's all for. Dan Dennett, the philosopher, always asked the question "cui bono?", or "who benefits?" In biology the genes benefit, and in culture the memes benefit. As I hinted at earlier, this is a radically different view from standard cultural evolution theory. I think standard cultural evolution theory can deal very well with most of human history and prehistory, but in the present day I don't think it can. I think now we're seeing dramatically the effects of meme competition on people's lives and minds; the difficulties we have with fake news, the awful spreading of violent and horrible videos of all sorts.

You can see the competition there between truth and falsehood, between nice things that make people happy and things that make them miserable, and the stress we are under by the fact that our brains are being used by all this information for its own sake, to get itself carried on. So I am contemplating writing more now about that and about what I think is happening next, which is that the information is itself taking over the processing and storage power of the technology we're providing. We think we're providing it for ourselves, but I think it's actually being taken over now by a third replicator. If that's what comes out of memetics and helps us to understand the problems we have with AI and other things at the moment, that will be really useful. And if I'm completely wrong, people can go on ignoring it.

The vision of reality you present in *The Meme Machine* and develop in *Consciousness* is a quite sobering one. In this vision we are merely the passive carriers of memes like the people in "*Matrix*" reduced to the role of energy sources. Don't you think that highly pessimistic portrayal has contributed to the declining interest in conception?

I don't know. When you say it's sobering you could be referring to quite a lot of things. Let's take a few of them. One thing is free will. It seems absolutely obvious to me that we can't have free will, at least in the sense that most people think about their own free will, meaning we feel as if we can have a thought, and quite independently of anything else we can enact that thought, I can do anything I want just because of my thought. Yet you only need a cursory understanding of the brain to see that doesn't make any sense. The precursors of every action this body does are in the brain and in the rest of the body, and it's those things, the environment around me and the history of my life, that determine every action that I take. If by "sobering" you mean going from being drunk and happy to being sober and looking at things straight then yes, I agree. Is it horrible? Many people say "it's so depressing, how could I live without believing in free will?" The answer is, I suppose, a bit like what we were saying earlier about changing your mind about scientific theories. It's hard to give up the natural impression we all have that we are free in our minds, that it's me and my mind controlling the brain. It's not, but once you get used to that it's actually wonderful because you can look at what's driving the behavior of your own mouth when it speaks, and you can become more critical and careful about it, more amazed by the wonder of it. It's tricky, it's hard work to give up believing in free will, and you have to be morally cautious as well. But it's wonderfully pleasing in the end.

Now let's take the memetic view. If you take the view that all of our culture is selfish memes that have got here by being copied, then you begin to understand what it means to pass on different memes. You think about your own behavior in terms of which things you pass on and what the consequences are. I think about my little grandchildren, watching them learning language, learning to draw and other things like that. It's so wonderful, and I'm thinking about the memetic environment that they're growing up in, so I agree it's sobering but I don't think it's depressing. But if we move on to my idea about technological memes, or "tremes" as I have come to call them, that's worrying. My feeling is if I'm right and the machinery is increasingly being taken over by algorithms and structures of information that we can't even see because they're evolving in their own way for their own sake, then we are doomed to a scary future if we don't work out what's going on and react appropriately. That's why I'm interested in talking about what I think is happening, which is just another example of the power of Darwin's great insight into where design and creativity come from.

The conception of memetics is inextricably linked with considerations on the possibility of re-programming our mind to make it a carrier of only those memes that we ourselves wish to serve. Isn't the thought that we could consciously become carriers of selected memes equally absurd as the conviction that we could decide which genes we wish to pass along to our offspring?

Yes, but the situation is very different with genes. It takes the technology of gene editing and so on for us to do that. But in terms of memes I think we need to see ourselves as meme machines. That was the whole point of the book, that we are both gene machines and meme machines; the genes have given us a tendency to like some things and not others, to pass on certain things and not others, to believe certain things and not others, to get trapped by certain ideas and not others, and so on. We start with the genetic basis of what we're like, then we become educated: we become able to understand language, to read and write and everything else, and we have to become selective imitation devices. That's what I've called us, selective imitation machines. It's our job to survive, to select memes. So what you're describing is what we do anyway, and what we have always done since humans existed. A kind of a battle to choose the memes that we believe will be good or true or helpful or useful to us in some way, and to reject the ones that won't. The problem is that we're not very good at it. And this is why I described a lot of the tricks that different memes use. The most obvious example here is religion.

I'm just horrified at the world today considering how, when I was a student 50 years ago, I honestly believed that religion was just going away. If anyone had told me then that there would be religious-based wars in the 21st century I would have been shocked. The point is that this is relates to the meme tricks. Religions have developed what I call the altruism trick, persuading their followers that they are good people if they believe this religion. This comes up again and again in all the monotheistic religions, the idea that you're good because you believe in Jesus or because you follow Allah, or you're good because you give money to the temple or whatever it might be. That is a horrible trick to play because our genes have given us the desire to be liked, to be respected, because if you're liked and respected, you have a better time in life, you get more stuff if you're seen to be good. We want to be seen to be good, even if we're not. Religions just jump on that bandwagon and make people feel good by belonging, and yet we know the awful harm that religions do both to people and societies. Those are the kind of venues in which this competition plays out, and if I understood your question correctly, we have to be self-programming and hope that we have friends, colleagues and others who can help us with this job, to think critically about the effect of the memes that are all around us all the time, which ones we pick up, and in particular which ones we pass on to other people. But it gets harder and harder the more memes there are.

Yes, it's a difficult task, so I'd like to ask you about what you do to manage it. As a sort of antidote to the dictatorship of memes, you suggest cleansing the mind through meditation and concentration on the here and now. How can you be certain that the belief in

immersing oneself in the here and now isn't just another memeplex that you yourself are the carrier of?

Oh, it is just another memeplex, absolutely. But I don't think I've ever advocated for it in the way that you're implying in the question. I don't go around saying "oh, everybody must meditate." I began meditating back in the early 1970s. I've been meditating every day now for more than 30 years. While I can't prove it, I feel that this has helped me enormously in my life, to be less angry, to be less greedy, to be less aggressive, all the things I have a great tendency toward. But I can't prove that, because it could be just due to getting older. What I have written about and stand by is that most forms of meditation are a kind of meme weeding. This applies particularly to the kind of Zen meditation I do—*zazen*, which means "just sitting," and it is precisely that—just sitting.

I like the analogy of clearing, or weeding away, the memes because I'm also a great gardener. I spend a lot of time planting trees, growing vegetables, looking after the chickens, and I have a huge garden that is always full of weeds. So the obvious analogy came to me. When you've spent a long time getting rid of all the weeds from a patch of ground, it's all clear and ready to sow whatever it is you want to grow there, but if you don't do that within a couple of weeks it will be all covered in green again, because there are seeds everywhere that jump in at any opportunity. Learning to meditate is very much like that early stage, when you haven't had much practice your brain is not used to dropping into quiet states. It's really just brain training. What happens is, you sit down and try to calm your mind, then a bit of space clears and another thought comes in, followed by another and another. As soon as there's space, another one turns up. That's what the world is like in our minds, or when we turn on our computers. Another email comes in, another WhatsApp message, another text. That was the analogy that I made. I stand by that analogy, but I am not saying to people "oh, you have to go out there and do meditation in order to weed your mind of all its memes." You can't do that. You are a meme machine; you are a creature of memes. You can be selective about those memes, and if you choose to, you can take on the meditation memeplex. Sit down every day, quiet your mind and ultimately get to where your mind will just quiet itself when you sit down in meditation because it's had years of practice. I think it's helpful, but it's not for everyone, and there are dangers in meditation. It's not as if you can just clear your mind and start again. All I can do is choose and choose wisely between the different memeplexes that are around, and that is hard.

In *The Meme Machine* you stated rather succinctly that "Today's psychotherapy is a kind of memetic engineering, but it is not based on sound memetic principles." Could you perhaps elaborate on the subject of understanding psychotherapy through the prism of memetics?

There are so many different kinds of psychotherapy, from psychoanalysis based in ludicrous theories from a hundred years ago to modern kinds of CBT, which work pretty well for certain things. And there are all kinds of spiritual therapies and other things. But almost all of them are based on the concept of working with a self, and that the self is something that is important, to be made as good and as happy and as content as possible, to deal with our problems, and so on. In the memetic view, the self is a construct of memes. We only have this illusion of there being a "me" inside.

Let's go back to the self for a little longer. I think the way we need to think about the self now—and I'm absolutely not alone, there are lots of neuroscientists and philosophers writing books about this for the last 10 years or so—is that the self is a construct of the brain. The self, the feeling of I, the responsibilities of being myself and so on, are all constructs of the brain. This leads us to be dualists, to believe that we are separate from the world that we're in. We are sure that we are a conscious self in here looking out through our eyes and there's an external world out there. I reject that, as many other people do, and my view of memetics replaces the dualist view. I replace it with the idea that one of the reasons why the self is so powerful in a person's mind is because of the memes that have clung to the idea of self. They have built up the identity of that person, and cause them to cling to the theories they believe in. I don't know of any memetics therapy that might exist, because I'm not a therapist and I'm not trained in therapy. There may be some out there, but the vast majority don't take that view of the nature of self. And this actually relates to a fascinating question in meditation. People talking about meditation and progress in meditation often say that you need a really strong self before you can give up the self. For example, the Buddhist idea of self is something ephemeral that comes and goes, not permanent in any way. That argument has played out for over 2000 years. And then there's this idea of therapy versus enlightenment. If enlightenment is in some ways giving up that false sense of self, should you go into therapy and obliterate people's self? You can't do that, because they won't function as people in the world. So there is an alternative argument that you need to create a strong and stable self before you can go through the processes of acknowledging that self is just an ephemeral thing that comes and goes. I don't have any answers, but that's some of the context in which I said long ago that therapy is not based on memetic principles. If it were, it would have to tackle that difficult issue about the nature of self, about coming to terms with yourself being ephemeral and non-persistent, and there are plenty of therapists arguing about this kind of thing. But more obviously, you would look at the memes that people have been infected with throughout their life, the ones they cling to and spread most enthusiastically, and think about how you could help that person to live their life better. If they were taking in different kinds of memes, if they got themselves in different sorts of circumstances that they might meet different kind of memes, you would have them think about how they propagate the memes and the effect that has on their relationships with other people. This would not be dramatically different. You'd still be doing many of the same things, but it would be a different way of thinking about therapy. But as I said, I'm not a therapist, so I have not pursued that line of inquiry, nor tried to create any kind of therapy.

Do you think that different modalities, therapeutic modalities are different memeplexes?

Yes, of course they are. Take Rogerian theory, Jungian analysis, CBT or transpersonal therapy. They are indeed big memeplexes, and one's hope is that by bringing such a memeplex to somebody in trouble you can help them, that the memeplex will help them sort out their primary problems. There's a fundamental difficulty here. It sounds like I'm saying everything is a meme. So I need to clarify that all of culture is a meme. Everything I have ever said is a meme, every speech that you have ever heard is a meme, any information that is copied from person to person or person to book or person to computer or whatever is by definition a meme.

That's what a meme is, that kind of information. That doesn't mean that everything is a meme, that all our thoughts are memes. If you don't send something on it never becomes a meme. All the skills you have, physical skills like walking, riding a bike, driving a car; gardening is a wonderful example, memes are involved, but the skills of actually dealing with the earth and the plants are something else. These things are not memes; they are skills that you've learned as an animal with learning abilities. But all scientific theories, all religions, all political theories, money—they're all memes because they have been copied and passed on.

Do you think that psychotherapy employs the kinds of tricks used by memes, similar to religions?

Yes, to some extent. I think that would certainly be true of psychoanalysis, which as a memeplex jibes very well with people's dualistic views, the importance of their self, their worries about their parents and all such things, and some people love it. Some people are willing to spend enormous amounts of money for what we know has very poor outcomes. You can have 10 years of weekly psychoanalysis and not be much happier or function better in life than at the start. Some people find it miraculous, but in general the evidence suggests it's not great. So it's playing those kind of tricks because it appeals to people, because it seems to make a lot of sense, and then the evidence shows that most of Freud's theories are incorrect. There are little bits in there with some value, but most of it is wrong, so that would be an example. My knowledge of modern therapies is limited, but I can see among the crazier versions, which I come across through parapsychology, that they prey on the same longing for the self to be important and feel better. They use similar tricks to religions in some ways. I think one of the most worrying ones is the idea that mixes spirituality with therapy in a very unhelpful way. "You will be a spiritual person. You can be more spiritual if you do these things and if you have this therapy, which will raise your vibrations to a higher spiritual level." That sort of really wacky thing uses very unpleasant tricks because spirituality sounds so wonderful, and if you can get it by going to a particular therapist, some people will be drawn in regardless of evidence.

In the end it comes down to evidence. When you look at the evidence for different kinds of therapies, one of the most interesting things to come out in recent years was that most types of psychotherapy have some good effect, which occurs within the first one or two sessions. But that effect is not dependent upon the type of therapy, the individual person, or the length of time that therapist has been practicing. In other words, it looks as if a large proportion of psychotherapy is operating through the perfectly normal human interactions of sitting and listening to people. It's not helped by years of training in different therapy methods. The evidence is more complicated than that, and since that global finding it's been discovered that some therapies actually do some harm, the majority just do a little bit of good, and CBT seems to do pretty well for certain things. But I keep going back to evidence. It's evidence that made me give up parapsychology. It's evidence that has changed my mind about consciousness so many times, and that's what I would say about therapy. We need evidence that it works. What do you want from this therapy? Can we measure that change at the end? And if it doesn't help, that's an awful lot of money people are throwing at it.

And their priceless time. In studying your conceptions, while also investigating various abuses of science, I am fighting the urge to conclude that science is nothing more than another memeplex whose sole objective is the replication of memes. It is served exceedingly well by the citation index and all sorts of influence rankings.

Yes, I absolutely agree with you. All science is memes. The difference between science and religion, or science and popular beliefs of various kinds, is evidence, and that's what training as a scientist entails: how to assess evidence, how to find evidence, how to do experiments and so on. That's what makes it different, that's what has given us all the technology that we have now and what has helped us understand the world. Of course, it's a never-ending process to try to understand it. It's interesting that you raised the citation issue, because I don't think it is the best science that gets the most citations. Nevertheless, it is a way of finding out which scientific contributions are most valued by other scientists. Because the more they're cited the more other people appreciate them. But what worries me is the effect is a bit like Amazon books or so many other modern things. Once you have an index like that, people will cite the most popular ones because other people have cited them, and it exaggerates the difference between the uncited and the very frequently cited. I don't think that's a good thing.

Does this mean that the life of an idea in science, particularly psychology, subject to the dictates of fashion rather than other, rational laws?

Obviously many meme problems will occur: fashions, what people like and the things that people want to be true are probably more often cited than the ones they really wish weren't true. But in the end the scientific method is about taking the evidence seriously, and that's what we should all be doing. I think it's becoming very hard now, considering the way universities are so stretched with too many students and all the other problems that we have in academia. Nevertheless, science has always had to deal with people who don't like the things that fine science discovers, and we have very good methods for pushing back. That's what science is all about. And we're just working in a different way now than we did 50, 100 or 200 years ago, when there were so few ideas around. When Newton came up with his laws of motion, for example, it was very difficult to get any new ideas out at all, but people wrote letters to each other. Gradually the ideas got out and everybody heard about them. There weren't so many ideas to compete with in those days. Now, any idea you put out has to compete, and that makes different problems in terms of popularity.

What in your opinion are other that mentioned before, big challenges presently facing psychology?

Probably the replication crisis. But I think psychology has long had a problem that people can't identify. When I started studying psychology, there wasn't any neuroscience, there wasn't any cognitive psychology, all that sort of started when I was a student. And when I say I'm a psychologist, it's those kinds of things that I'm talking about, while for the public, if I say "I'm a psychologist" they say "Oh, should I lay down on the couch?" That's not what I do. So we have this problem in psychology, which is breaking apart in a way, and I think this needs to happen. I think we need to discriminate between the neuroscientific side, which is based on what's happening in brain, and the therapeutic side, which is aimed at actually helping people. I'm not sure where things are going, but there's a crisis of identity that's long been brewing in psychology.

Will the development of artificial intelligence help us find answers to the big questions posed by psychology?

Yes, and this really interests me. I think that most people are worrying about the wrong sort of AI. They are concerned either with robots with artificial intelligence in them, in other words, the actual entities that are intelligent, or they're worrying about artificial intelligence that we have created to do things. For example, the internet of things, and self-driving cars, which will have artificial intelligence driving them. These are, of course, genuine worries and people are dealing with them. What I'm concerned about is the idea that AI may be self-evolving as we create more and more servers, phones, computers and massive storage capacity using an increasing amount of the energy produced on this planet by fossil fuels. I suspect that this massive explosion of information is giving rise to a new replicator; information, algorithms and computer programs capable of copying other programs. We may not even be able to see everything that's going on in this new evolving sphere. If you think about it this way, we should ask how our intelligence came about? It came about by ordinary biological evolution. A multicellular organism with billions of brain cells in it, each one a living cell—all of those things coming together doing pretty dumb things that, when put into a body, amount to intelligence. There's no such thing called "intelligence" inside us. It's all the things that all these cells do which makes the behavior of a human being intelligent. I would say what's going on in the cloud in cyberspace is exactly the same process, bits here and there are all contributing dim things, but ultimately the result is intelligent. So the fact that I can talk to Siri or say "OK Google" and get an answer, that's intelligent. And if you think of all the processing going on out there, with millions and millions of cameras sending information into the cloud all the time, video and still pictures, endless audio, and the ability of the machines to control what we say, do, and think all the time, and to actually speak to us—that's intelligence, and it's evolving. It's evolving for its own sake, not for ours, and that is the bottom line for memetics. I think that's what we should be worrying about in terms of artificial intelligence.

Question for dessert if you agree.

Sure.

I have a list of 30 questions that my readers would like to ask the most influential psychologists in the world. Please draw one.

27.

The most ridiculous idea in the history of psychology which was taken seriously?

What a really interesting question. I suggest two such ideas. The first one is that you could have a full and rich psychology without considering consciousness—or subjective experience. The second is the idea that you could treat a human or other animal as a "black box", measuring only inputs and outputs without any concern for what may be going on inside. Both were standard assumptions when I was a student but, happily, not any more.

Would you like to add something to our conversation or mention issues we haven't discussed yet?

I would like to give our readers a bit of encouragement and say that we have amazing brains and the capacity to think, and we live in an amazing world full of fantastic amounts of information. What could be more interesting than trying to understand the mind that is trying to understand itself? Psychology is a very special subject in that sense.

Selected Readings

Blackmore, S. (1982). *Beyond the body: An investigation into out-of-body experiences.* London: Heinemann.

Blackmore, S. (1993). *Dying to live: Science and the near death experience.* London: Grafton.

Blackmore, S. (1996). *In search of the light: The adventures of a parapsychologist.* Amherst, New York: Prometheus Books.

Blackmore, S. (1999). *The meme machine.* Oxford and New York: Oxford University Press.

Blackmore, S. (1999). Waking from the meme dream. In G. Watson, S. Batchelor, & G. Claxton (Eds.), *The psychology of awakening: Buddhism, science and our day-to-day lives* (pp. 112–122). London: Rider.

Blackmore, S. (2001). Evolution and memes: The human brain as a selective imitation device. *Cybernetics and Systems, 32,* 225–255.

Blackmore, S. (2005). *Conversations on consciousness.* Oxford: Oxford University Press.

Blackmore, S. (2010). Dangerous memes; or what the Pandorans let loose. In S. Dick & M. Lupisella (Eds.), *Cosmos and culture: Cultural evolution in a Cosmic Context* (pp. 297–318). Washington, DC: NASA.
Blackmore, S. (2011). *Zen and the Art of consciousness*. Oxford: Oneworld Publications.
Blackmore, S. (2012). Turning on the light to see how the darkness looks. In E. S. Kreitler & O. Maimon (Eds.), *Consciousness: Its nature and functions* (pp. 1–22). New York: Nova.
Blackmore, S. (2013). Living without free will. In G. Caruso (Ed.), *Exploring the illusion of free will and moral responsibility* (pp. 161–175). New York: Lexington Books.
Blackmore, S. (2016). Delusions of consciousness. *Journal of Consciousness Studies, 23*(11–12), 52–64. Also reprinted in *Illusionism as a theory of consciousness*, Ed. Keith Frankish, 2017.
Blackmore, S. (2017). *Seeing myself: The new science of out-of-body experiences*. London: Robinson.
Blackmore, S., & Hart-Davis, A. (1995). *Test your psychic powers*. London: Thorsons.
Blackmore, S., & Troscianko, E. (2018). *Consciousness: An introduction* (3rd ed.). London: Routledge.

References

Blackmore, S. (1987). The elusive open mind. *Skeptical Inquirer, 11*(3), 125–135.
Blackmore, S. (1999). *The meme machine* (2nd ed.). Oxford and New York: Oxford University Press.
Dawkins, R. (1989). *The selfish gene* (2nd ed.). Oxford: Oxford University Press.
Dennett, D. C. (1995). *Darwin's dangerous idea: Evolution and the meaning of life*. New York: Simon & Schuster.
Henrich, J. P. (2016). *The secret of our success: How culture is driving human evolution, domesticating our species, and making us smarter*. Princeton: Princeton University Press.

9

Joseph E. LeDoux: Fear, Anxiety, Emotions, Consciousness and Evolution

Start by asking what you want to understand about behavior, mind, and/or brain rather than by choosing a method you want to acquire. The questions are eternal but the methods are fleeting.

Joseph LeDoux photo by Diemut Strebe

T. Witkowski, *Shaping Psychology*,
https://doi.org/10.1007/978-3-030-50003-0_9

At the beginning of *The Emotional Brain*, Joseph LeDoux writes:

> My father was a butcher. I spent much of my childhood surrounded by beef. At an early age, I learned what the inside of a cow looks like. And the part that interested me the most was the slimy, wiggly, wrinkled brain. Now, many years later, I spend my days, and some nights, trying to figure out how brains work. And what I've wanted to know most about brains is how they make emotions. (LeDoux 1996)

It is indeed a quite unusual beginning for someone who came to revolutionize the way psychologists understand emotions.

LeDoux began his research under the wing of Michael Gazzaniga, a leading researcher in cognitive neuroscience. They began working together on understanding the functioning of split-brain patients. To further those efforts, they built a laboratory inside a trailer hitched to a pumpkin-colored Ford van and frequently traveled from Long Island to see patients at their homes in Vermont and New Hampshire.

As a postdoctoral fellow, he received technical training in state-of-the-art neuroscience techniques. Seeing that techniques for studying the human brain were limited at the time, he turned to studies of rodents where the brain could be examined in detail. He chose to focus on a simple behavioral model, Pavlovian fear conditioning. This turned out to be an outstanding choice, as it allowed him to discover the flow of information about a stimulus through the brain as it comes to control behavioral responses by way of sensory pathways to the amygdala. His discovery gave rise to the notion of two sensory roads to the amygdala, with the "low road" being a quick and dirty subcortical pathway for rapid activity behavioral responses to threats, and the "high road" providing slower but highly processed cortical information. His work shed light on how the brain detects and responds to threats, and how memories about such experiences are formed and stored through cellular, synaptic and molecular changes in the amygdala.

LeDoux took advantage of a wealth of new opportunities to cooperate with numerous curious and insightful researchers focused on the secrets of the brain, leading to more new discoveries. Studies conducted with Maria Morgan in the 1990s implicated the medial prefrontal cortex in the extinction of responses to threats and paved the way for understanding how exposure therapy reduces threat reactions in people with anxiety by way of interactions between the medial prefrontal cortex and the amygdala (Morgan et al. 1993). In addition, years of collaboration with his New York University colleague Elizabeth Phelps demonstrated the validity of LeDoux's rodent work for understanding threat processing in the human brain (Phelps and LeDoux 2005). Studies done in cooperation with Karim Nader and Glenn Schafe triggered a wave of interest in the topic of memory reconsolidation, a process by which memories become labile and subject to change after being retrieved (Nader et al. 2000). This led to the idea that trauma-related cues might be weakened in humans by blocking reconsolidation.

In 2012, LeDoux published an article discussing emotional brain functions in animals and humans titled "Rethinking the emotional brain," in which he argued that the use of mental state terms that are derived from human subjective experience should be restricted to mental states. The common practice of calling brain circuits that detect threats and generate behavioral and physiological response to threats "fear circuits" implies that these circuits are responsible for feelings of fear. LeDoux, however, argues that so-called Pavlovian fear conditioning should be renamed Pavlovian threat conditioning, to avoid the implication that "fear" is being measured in rats or humans. Threat conditioning may condition "fear" in humans, but most studies only measure behavioral and/or physiological responses, which is all you can measure in animals.

The results of LeDoux's scientific inquiry on amygdala processing of threats have helped us to understand exaggerated responses to threats in anxiety disorders in humans. The conclusions from these studies and their practical implications are described in his 2015 book *Anxious: Using the Brain to Understand and Treat Fear and Anxiety* (LeDoux 2015). In *Anxious*, LeDoux highlighted the notion of survival functions mediated by survival circuits, the purpose of which is to keep organisms alive rather than to generate emotions. In one example he discusses, defensive survival circuits exist to detect and respond to threats and can be present in all organisms. However, only organisms that can be conscious of their own brain's activities can feel fear. Fear is a conscious experience that occurs the same way as any other kind of conscious experience: via cortical circuits that allow attention to certain forms of brain activity. LeDoux argues that the only differences between an emotional and non-emotional state of consciousness are the underlying neural ingredients that contribute to the state. His latest book, *The Deep History of Ourselves* (LeDoux 2019), explores the ancient roots of survival behaviors (e.g., tracing the origin of defense and feeding to single cells living billions of years ago), and argues that consciousness of the types humans have may have arisen in our early hominid ancestors.

Professor LeDoux, the results of your research have helped to answer many of the fundamental questions posed in psychology. One of them was the question about relations between emotions and cognition, which during the last decades has sparked a number of passionate debates. Perhaps the most important adversaries in these discussions were Robert Zajonc and Richard Lazarus (Zajonc 1980; **Lazarus** 1984)**. In retrospect, do you consider this dispute resolved? Or maybe the problem of "what comes first, emotion or cognition?" was simply phrased incorrectly?**

With time on your side, you sometimes start to see things differently. My early work showing that a threat could activate the amygdala by way of subcortical sensory pathways, and thus at the same time as sensory cortex receives the signal, and before higher-order cortical areas did, suggested that affective processing could occur unconsciously. This aligned me with Zajonc—the affect (associated with the amygdala) could happen independent of cognition (associated with the cortex). But there was an aspect of Zajonc's view that bothered me. He said that affect was automatic, by which he meant it did not involve information processing. He equated with information processing with cognition, and concluded that affect could not involve cognition. But even mere sensory transmission by the retina involves information processing. I think he should have emphasized the difference between the information processing underlying affect and cognition, rather than trying to eliminate information processing from affect.

For reasons involving the work I did as a Ph.D. student on split brain patients, I thought of consciousness, including emotional consciousness in terms of cognitive processes that interpret situations in which we find ourselves. When I started doing emotion research I used a distinction popular in memory (explicit vs implicit) to characterize emotion—explicit emotion was the conscious feeling and implicit emotion involved circuits that control behavioral and physiological responses. So implicit fear was a product of the amygdala and explicit conscious fear of cortical circuits. But the amygdala simply became a fear center without any distinction between explicit and implicit. This was true of scientists and laypeople alike, and I was also sloppy myself at times.

That's why stopped using fear to refer to implicit processes. By restricting fear to the experience, I restrict it to a cognitive interpretation. I think Zajonc and I were right about non-conscious factors that can affect how we respond to "emotional" stimuli, I was wrong to call the nonconscious processing circuits emotional circuits, and he was wrong to deny that information processing was involved.

Unlike for memory, in the study of emotion the implicit-explicit distinction did not gain traction. It's a little odd since Freud had pave the way for this notion, but perhaps antipathy for Freud was part of the problem. More generally though the problem was related to the lingering effects of behaviorism, which eliminated mental states as causes of behavior but retained mental states as names for intervening variables that connect stimuli to responses. When behaviorists became physiological psychologists, and brain areas and circuits came to be physical instantiation of intervening variables, fear was said to control behavior. These were "psychological" (they controlled meaningful behavior) but were not "subjective" (they were not "experiences"—more like what I meant by implicit emotion).

This nonsubjective fear came to be the explanation of fear disorders and the target of treatments. But the treatments are not as effective as sufferers would like. They may be less timid and avoidant, and less hyper-aroused, but still feel fearful and/or anxious. That makes sense if all you've done is changed the circuit that controls behavioral and physiological responses.

And that's why I have been trying to rethink constructs (fear, emotion) that are stuck in 1940s psychology by putting them in terms that are more consistent with what we know about mind and brain today.

The results of your latest research, including what you described in *Anxious*, today provide us with answers not only to questions about the brain's emotional and defense mechanisms, but also to serious philosophical questions, such as the essence of memory, consciousness, and the mutual relationship between them, while your latest book, *The Deep History of Ourselves*, also explores evolution. When formulating your research questions, were you looking for answers to these big questions, or are they the result of simpler research questions and the extensive collection of data?

Deep History is about the evolution of behavior in the context of life, and especially the evolution of memory, emotion, self, and consciousness. Most of my books have been about these topics: memory, emotion, self, and consciousness. I was fortunate to have done my Ph.D. studying split-brain patients, which hooked me on these big questions. But I had to make a living as a young scientist and turned to a focus point—emotion—which was understudied at the time in neuroscience. I felt I could ask simply neurobiological question about emotion were relevant to the big questions, but without having to try to apply for a grant about big questions, since the latter would never have been funded. Even putting "emotion" in the title of my first grant got it rejected. I change "emotion" to "emotional conditioning" and that satisfied the lingering behaviorists who dominated the field. That grant got funded over and over for more than 30 years. During that time, I satisfied my urge to think about the big questions by writing books. *Deep History* ties it all together with two radical theses at the end: (1) contrary to popular ideas such as the limbic system theory, mammalian emotion did not pave the way for human cognition. Instead, cognition preceded emotion in evolution—that is, cognition was a precursor to emotions; (2) emotion were not initially products of natural selection in our mammalian ancestors but instead emerged an exaptation (Gould and Vrba 1982) that was made possible by other early human adaptations and/or exaptations such as: hierarchical cognition; mental schema; mental time travel; awareness of self as a subject (not object); and language (not speech but the cognitive underpinnings of language that all mental modeling to parse social situations into who is going to do what to who and when, all in a fraction of a second). Emotions are cognitions.

The circuits that control behavioral and physiological responses are not emotional circuits; they are better thought of as survival circuits. Survival circuits occur in some emotions (those often called basic emotions) but not in others (secondary or social emotions). The only thing that allows both to be in the same category is that in each case there is an awareness that something of biological or psychological significance is happening to one's self. No self, not fear, and no other emotions.

You have repeatedly drawn attention to the difficulty caused in psychological research by the presence of colloquial psychological concepts. Could you elaborate on this by explaining what the primary problem is and how we can avoid making mistakes in interpreting observed phenomena?

There's a long and complicated literature on folk psychology. A popular idea is that all these lay terms we use in day to day life will be replaced by scientific language. Examples are given from physics and chemistry that show how this will be done, in terms of brain processes. But as George Mandler and William Kessen (1959) pointed out, psychology is not the same as other sciences. The reason is, because "man studies himself." The social psychologists George Kelly (1955) offers a useful addendum. Our vernacular language should be replaced for aspects of psychology, but is needed for other aspects. We need it he says for our subjective experiences. Other aspects that more automatic processes can and should be replaced. In terms discussed above, neuroscience terms (genes, molecules, cells, synapses, circuits) are the way that defensive responses based on innate wiring or conditioning should be described. But we still need mental state words to name mental states. That is not dualism in the philosophical sense. It is simply saying that the systems that underlie the complex cognitive processes that make mental states are not of the same category as those that control behavioral and physiological responses. Folk psychological terms have a place in understanding mental life but not in understanding more forms of behavioral control. By the same token, this is why changing the more primitive systems is not the way to change how one feels about their life. It might help some but is not the way to change the content of experience, which is only understandable by the interaction of personal, social, and cultural factors.

Does the counterintuitive distinction between survival circuits and emotions resulting from your research have a chance at replacing the way we currently think about emotions such as fear, anxiety, etc.?

Any new idea has faces tremendous obstacles. It makes sense to me, as I think it can clarify some very confused ideas about what emotions are. Time will tell.

The results of your research are of fundamental importance not only to theory, but also to psychotherapy, as you have shown, among other places, in *Anxious*. To what extent are they employed by contemporary therapists?

It's a process. I find therapists open to my ideas, since current methods are far from ideal. And in general they embrace my claim that as long as subjective experience is marginalized as an end point therapy will never achieve its goal of improving well-being, which is all about how you experience your life. Practicing clinicians are more open than research psychiatrists who are locked in by an emphasis on highly reductionistic explanations (genes and molecules) when experience is a emergent factor involving complex interactions between many low level factors, and a number of high level ones (self, social organization, culture).

There is a good deal of enthusiasm today for circuit explanations. I share this enthusiasm, but there is a problem. You have to know what you are looking for psychologically to find the circuit. If you think of fear in terms of defense circuits you will never figures out how to make people less fearful. Fear is a cognitive state of being in harm's way. It too involves a circuit but not simply the defense circuit. The latter is part of but not the defining part of the circuit that underlies subjective fearful experiences.

Looking through the prism of your knowledge, you are certainly able to distinguish therapeutic methods that are incompatible with how our brain works. Which of those methods do you consider particularly dangerous?

I'm not sure that's true. To the extent that a procedure has some effect it does so via the brain. What is a "dangerous therapy" for one patient may be useful to another. I say this but am not a therapist so I don't say it with authority.

That said, if we simplify (maybe oversimplify) things, it is pretty clear that the circuits underlying the control of behavior maladaptive behaviors (excessive freezing and avoidance) and hyper-arousal (of body and brain) are different from those that allow you to experience a mental state of fear or anxiety, or yourself as a fearful or anxious person. These are the responses I focus on in trying to relate therapy to the brain. Given that you might think that I would then say CBT is the way to extinguish behavioral and physiological responses since CBT generally includes a component by which people are exposed to threatening cues to extinguish their potency in eliciting behavioral and physiological responses. But the fact is that CBT involves much more, and exposure inducing extinction is not always the main focus. And often exposure focuses on cognitive extinction, which involves somewhat different processes than the mere stimulus repetition that underlies laboratory studies of extinction. So when we equate what we know about extinction in the brain from lab studies with what goes on in CBT we often are comparing apples and oranges. If we are going to use brain research to help understand therapy we need to be more rigorous in how we relate specific aspects of therapy to the simple models used in the lab.

When *The Emotional Brain* came out, it provided arguments to opponents of memory recovery therapy, showing that reconstruction of repressed memories is highly unlikely This was an extremely important contribution to the discussion of a social problem that suddenly became the focus of much attention. I wonder how you see these problems today from the perspective of the years that have passed since the publication of this book. Do we have more facts at our disposal that allow us to formulate a more unambiguous position on the legitimacy of using the notion of repression and work on recovering memories?

In *The Emotional Brain* I said that there was an alternative explanation to repression—stress hormone-induced amnesia. I don't think I dismissed repression outright but probably said there was not much empirical evidence for it. These days though Michael Anderson and others study repression using cognitive science (Anderson 2006). Also, there is work on directed forgetting. So while Freud is no longer revered to the same extent he once was, that doesn't mean everything he said was wrong. We may need to be more open-minded about things that we reject for non-scientific reasons (one's opinion of Freud the man), or because the science didn't exist when the ideas were rejected.

There are also memory modulation approaches like blocking reconsolidation. Work we did in rats and people in the lab stimulated a lot of clinical interest in this. But it has yet to emerge as a practical way to help people clinically (Phelps and Hofmann 2019).

But I meant not so much the process of repression itself but the possibility of reconstructing repressed memories with the help of a therapist. Is this process possible in the light of the latest research results?

I think, in principle, this reconsolidation can be used to help people. It's just a matter of finding the proper way to implement it in a complex brain, such as that of a human. Efforts have been disappointing (Schiller and Phelps 2011), but target article in *Behavioral and Brain Sciences* involving clinicians and scientists offered a more optimistic view (Lane et al. 2015). Time will tell.

This topic is also linked with the question of the credibility of witness statements, which your research has also shed much light on. Today, after several decades of intensive research into memory and remembrance processes, can we unambiguously determine their value?

I have never worked on this. But my work, especially on reconsolidation, has been applied to it. Nevertheless, it is important to point out that most of the lab work on reconsolidation in animal and humans involves implicit conditioned memories (in other words conditioned trigger cues) not conscious memories. Eye witness testimony is all about conscious memory. There is a bit of evidence for explicit memory reconsolidation in humans but we have to be careful not to generalize findings from conditioning to conscious memory—different brain circuits are involved. Regardless of how reconsolidation fits into these, eye witness testimony is known to be somewhat unreliable. But has traditionally been the only option. While video surveillance is maybe making it somewhat less important, new tools for facial substitution in video raise problems for its use. Human culture is complicated. We have to do the best we can with what we have. But ultimately we have a supervisory layer (a government) that we can trust to protect the public good. Right now that is disintegrating world-wide.

Why are scientific discoveries adapted by practitioners so slowly and with such difficulty?

In general I think takes a while for science to make its way into applications. That's a good thing. For example, if educational practices changed with every psychological or neuroscience fad before things are really understood, the system would be a mess. But more generally, educational and therapeutic procedures have to be approved by governing groups (State or City Departments of Education, or for therapy APA or other large, powerful organizations). The VET changes before blessing their use.

Your research, especially on memory reconsolidation or memory erasure, for some people raises ethical questions. I realize that from the point of view of a scientist, the study of processes and possibilities of memory modification is morally neutral. Doubts concern the possible use of these discoveries in practice. Are we allowed to use memory engineering, and if so, in what circumstances? What is your attitude to this question?

Well, every therapy session is a memory engineering event. But also ethical fears misunderstand the power of procedures that have been imposed. The concern is about whether conscious memory will be erased. But that is a misguided concern for two reasons. As I noted, most of the lab work on reconsolidation has involved implicit conditioned memories, not conscious memories. And second erasure is the wrong word. At its best, the procedures might dampen the impact, rather than remove the content, of a memory. That said, the mere act of retrieval can change content, either changing the strength of a memory (enhancing or weakening) it or replacing it with a new memory (a false memory). Reconsolidation may have something to do with these as well, but we should not assume that every effect of retrieval is due to reconsolidation.

I think these fears come from the fact that from time to time, psychologists engage in ethically questionable activities just as they did at the beginning of the twenty-first century, when psychologists affiliated with the APA were involved in work on techniques for

interrogations to be applied as part of the war on terror. This is not the first time that people from our discipline took active part in researching and developing methods used against other people. We can recall the example of the CIA-inspired MKUltra program. Are you not afraid that research on memory reconsolidation could be used for a similar purposes?

New discoveries, by virtue of being new, can offer opportunities for doing good and harm. That's why we have laws to regulate how people are treated. But there are several things to keep in mind. First, psychotherapy is a learning process, in which implicit and explicit memories are changed. Reconsolidation is just another potential way to do that. Second, a lot of the concern is about memory erasure. But the actual effect is more about updating rather than erasing. Third, the concern is also about explicit, conscious memory. The animal work has mostly focused on implicit memory circuits that control behavior. This is likely to be possible in humans. If someone shows powerful effects that "erase" human memory then the issue would need to be considered seriously. But I think the most likely result will be the ability to modify behavior and physiological arousal, and perhaps some modulation of the intensity of explicit memory. This is not so different from what therapy does now.

Psychoanalysts, especially those who call themselves neuropsychoanalysts, often refer to the results of your research, and you yourself for some time served on the editorial board of the journal *Neuropsychoanalysis*. What is your view on psychoanalysis today? Do you find potential for future research in it?

Freud was brilliant and we owe a lot to him, even if there are problems with some of his ideas. And I found neuropsychoanalysis interesting at first. It seemed truly about integrating neuroscience and psychotherapy. But it quickly came to focused on a narrow aspect of the interaction, adopting the idea of Jaak Pankseep as their foundation. His view of emotions were quite different from mine. So I stopped being an active participant in that effort. But from time to time I interact with analysts. I learn from them and they from me. I am open to lots of ideas and don't hold it against current analysts that not everything Freud said is correct.

Your research results have shaped our understanding of emotions and are the foundation for further investigations. Which of them do you consider the most important in the development of science, and why?

My Ph.D. thesis work on split-brain patients had a phenomenal impact on the way I think about mind and brain. Many in the field don't know that this was part of my past. But to the extent that the research I did later on the brain mechanisms of Pavlovian conditioning had an impact on the field, that would not have happened without my split-brain work. It gave me a hunger for the big questions about how conscious and nonconscious processes interact. Because ways to study the human brain were limited in the late 1970s, I decided to focus on how nonconscious systems work in animals.

From time to time I have been credited with implicating the amygdala in fear. That is not the case. Others had shown that amygdala damage in animals disrupted "fear" responses. What I showed was how a dangerous stimulus activates the amygdala to control the responses. That was a different approach. Most work was asking, "what does brain area X do?" But the split-brain logic taught me to treat the brain as a switching board that connects stimuli to responses, and new pathway tracing methods gave me a way to follow the logic. There were three of us working on this. Bruce Kapp, Michael Davis, and me. Between 1984 and 1990, we laid out the inputs and outputs. And that's what jump-started the Pavlovian "fear" conditioning field.

My 1996 book, *The Emotional Brain*, helped make the amygdala a household work in the study of emotion. I was talking about implicit emotion but that subtle point was lost and the amygdala not only became fear center but an emotion center.

My empirical work in the 1990s turned toward questions of plasticity and the molecular mechanisms involved. Much more work was being done on the hippocampus at the time using long-term potentiation to study plasticity spatial learning to study memory behaviorally. But it was hard to connect the two. Because my lab had identified the inputs to the amygdala involved in fear conditioning, we were to study both long-term potentiation and memory in the exact same circuits. This helped elevate the status of the amygdala as a model of learning and memory, not just emotion.

We were very involved in protein synthesis in the amygdala as the basis of memory formation. In 2000, Karim Nader, a postdoc in my lab, discovered that protein synthesis also determines the fate of memory after retrieval. This went agonists the dogma that memories are stabilized after learning and then retrieved later. The new work suggested that when memory is retrieved, it can be changed by a new round of protein synthesis that stabilizes a new memory. This ended up having a huge impact on the field, and is a major area of work today on memory, so-called reconsolidation, and on possible ways to use memory alterations to help treat psychopathology.

My more recent work has focused on the transition from reactions like freezing to actions like escape and avoidance. This kind of work was popular in the 1940s and 50s, and had big impact on the development of exposure therapy, but fell out of favor because of conceptual confusions. It was assumed that fear reduction motivates avoidance. But my view is that it is not about fear reduction, at least not in the sense we humans think of fear (as a conscious experience). To me, avoidance learning is implicit instrumental conditioning in with an implicit Pavlovian process weakens, providing negative reinforcement of the responses. Fear is a separate parallel process, as I mentioned above.

This avoidance work is not only relevant to fear and anxiety but also to addiction and eating disorders.

What I think about most these days goes back to the split-brain studies—how conscious experiences come about. The science of consciousness in thriving right now but is mostly about perception, as opposed to the kinds of experiences we have day to day. And fear is an excellent mental state with which to approach real-life consciousness. And, of course, work on fear itself has important implications for mental health.

I am, and certainly our readers also are interested in the issues you are presently engaging. What questions are you trying to find answers to these days?

I have three funding sources right now, and am using them to pursue the last two topics I mentioned above. One, from the National Institute of Drag Abuse, involves negative reinforcement—how a behavior that prevents a bad thing from happening is reinforced. The second grant is from the Templeton Foundation and is on defensive survival circuits and their relation to conscious fear experiences. In addition, I have funding from a private donor to continue to pursue the nature of anxiety and its treatment under the umbrella of what we call The Vulnerable Brain Project.

What are the big questions facing our discipline that we have yet to find answers to?

I believe that before we can make progress on big questions about mind and brain, we need to think more deeply about what the mind is and how it works. We think we know from our own experiences but that is an illusion. If it was that simple we wouldn't need science. Part of the problem is that neuroscientists get more training in how to use methods than in how to ask questions about the mind. We take too much for granted. At the end of his career, Pavlov penned a letter to young Russian scientists, urging them to not be content with the surface of things. Too often we are stuck at the surface based on conventional wisdom. We need to think deeper. Neuroscience training might benefit from classes in philosophical inquiry.

Someone with your achievements and experience can allow himself to engage in a bit of futurology. In your opinion, what will psychology look like in two or three decades?

I will answer for both psychology and neuroscience. But will talk about a hope rather than a prediction. In psychology and cognitive neuroscience, techniques such as fMRI and TMS are powerful research tools. In neuroscience, optogenetics has been a game changer. But the fact is, our ability to understand the brain is only as good as our understanding of the psychological process being investigated. Students often come to graduate school with a desired to learn these methods, but with less insight and passion about the question they want answer. Maybe we need to rethink training, and spend more learning to think critically about the psychological constructs that are used to motivate research and explain the data.

Perhaps our conversation will also be read by young people just getting their careers started. They are probably wondering what

specialization to explore. What areas of investigation in psychology are particularly promising? What should they be careful of? Can you offer a few tips?

Continuing my point in the previous answer, I would say start by asking what you want to understand about behavior, mind, and/or brain rather than by choosing a method you want to acquire. The questions are eternal but the methods are fleeting. Don't get me wrong, you do need to know how to do things, and it a good career move to be current in terms of techniques. But in parallel, ask yourself what you want to find out with those methods. I admit choosing a topic is complicated. But here's a suggestion. Survey the field thoroughly and make use you understand all its aspects. Don't just look at the field today. Learn the history of psychology and neuroscience. That can be very useful in seeing where present day constructs come from. It was my knowledge of history of "fear" in psychology that helped me see that our ideas on this were codified long ago and became unquestioned dogma. So if you find a topic you are passionate about unpack its dogma historically ("Why did they think of it that way" and "What were their ideas based on?"). Then rethink these with an open mind.

Selected Readings

Gazzaniga, M. S., & LeDoux, J. E. (1978). *The integrated mind*. New York: Plenum Press.

LeDoux, J. E. (1996). *The emotional brain.* New York: Simon and Schuster.

LeDoux, J. E. (2002). *Synaptic self*. New York: Viking.

LeDoux, J. E. (2015). *Anxious: Using the brain to understand and treat fear and anxiety*. New York: Viking.

LeDoux, J. E. (2019). *The deep history of ourselves: The four-billion-year story of how we got conscious brains.* New York: Viking.

LeDoux, J. E. (2020a, March 9). How does the non—Conscious become conscious? *Current Biology, 30*, R1–R4.

LeDoux, J. E. (2020b, June 8). Thoughtful feelings. *Current Biology, 30*, R1–R5.

LeDoux, J. E., Debiec, J., & Moss, H. (Eds.). (2003, September). *The self: From soul to brain* (Vol. 1001, pp. 295–304). New York: Annals of the New York Academy of Sciences.
LeDoux, J. E., & Hirst, W. (1986). *Mind and brain: Dialogues in cognitive neuroscience*. New York: Cambridge University Press.
Mobbs, D., & LeDoux, J. E. (Eds.). (2018). *Survival circuits: Current opinions in behavioral sciences*. New York: Elsevier.
Shiromani, P. J., Keane, T. M., & LeDoux, J. E. (Eds.). (2009). *Post-traumatic stress disorder: Basic science and clinical practice*. New York: Humana Press.

References

Anderson, M. C. (2006). Repression: A cognitive neuroscience approach. In M. Mancia (Ed.), *Psychoanalysis and neuroscience* (pp. 327–349). Milan: Springer.
Gould, S. J., & Vrba, E. S. (1982). Exaptation—A missing term in the science of form. *Paleobiology, 8*(1), 4–15.
Kelly, G. (1955). *The psychology of personal constructs* (Vol. I, II). New York: Norton.
Lazarus, R. (1984). On the primacy of cognition. *American Psychologist, 39*(2), 124–129.
Lane, R. D., Ryan, L., Nadel, L., & Greenberg, L. (2015). Memory reconsolidation, emotional arousal, and the process of change in psychotherapy: New insights from brain science. *Behavioral and Brain Sciences, 38*, 1–80.
LeDoux, J. E. (1996). *The emotional brain*. New York: Simon and Schuster.
LeDoux, J. E. (2012). Rethinking the emotional brain. *Neuron, 73*, 653–676.
LeDoux, J. E. (2015). *Anxious: Using the brain to understand and treat fear and anxiety*. New York: Viking.
LeDoux, J. E. (2019). *The deep history of ourselves: The four-billion-year story of how we got conscious brains*. New York: Viking.
Mandler, G., & Kessen, W. (1959). *The language of psychology*. Oxford, England: John Wiley.
Morgan, M. A., Romanski, L. M., & LeDoux, J. E. (1993). Extinction of emotional learning: Contribution of medial prefrontal cortex. *Neuroscience Letters, 163*, 109–113.

Nader, K., Schafe, G. E., & LeDoux, J. E. (2000). Fear memories require protein synthesis in the amygdala for reconsolidation after retrieval. *Nature, 406,* 722–726.

Phelps, E. A., & Hofmann, S. G. (2019). Memory editing from science fiction to clinical practice. *Nature, 572,* 43–50.

Phelps, E. A., & Ledoux, J. E. (2005). Contributions of the amygdala to emotion processing: From animal models to human behavior. *Neuron, 48*(2), 175–187.

Schiller, D., & Phelps, E. A. (2011). Does reconsolidation occur in humans? *Frontiers in Behavioral Neuroscience, 5*, 24.

Zajonc, R. (1980). Feeling and thinking: Preferences need no inferences. *American Psychologist, 35*(2), 151–175.

10

Noam Chomsky: Psycholinguistics

The study of the human mind, is one of the most difficult topics of science, and we don't need the Delphic Oracle to instruct us about its significance for human thought and life.

Noam Chomsky photo by Andrew Rusk, CC BY 2.0; https://creativecommons.org/licenses/by/2.0/; photo has been cropped

T. Witkowski, *Shaping Psychology*,
https://doi.org/10.1007/978-3-030-50003-0_10

It is more than difficult to write a short biographical note about a man who is considered to be one of the greatest living intellectuals and compared to such figures as Aristotle or Descartes. This task becomes even more difficult when we realize that we are dealing with someone who is responsible for breakthrough in linguistics, in spite of his claims that he is unfamiliar with many issues that a "true" linguist should be versed in, and therefore does not consider himself to be a truly professional linguist. Despite the fact that he never received an education in psychology, he has also made a breakthrough in psychology by becoming one of the leaders of cognitive rebellion in this field. Chomsky is also considered one of the most outstanding philosophers of the twentieth century, and his work in automata theory and his hierarchy have become well known in computer science. The 1984 Nobel Prize laureate in Medicine and Physiology, Niels Kaj Jerne, used Chomsky's generative model to explain the human immune system, equating components of generative grammar with various features of protein structures. His theory of generative grammar has also carried over into music theory and analysis. However, public opinion recognizes him primarily as a political activist and social critic.

Aged 16, Chomsky embarked on a general program of study at the University of Pennsylvania, where he explored philosophy, logic and languages. Frustrated with his experiences at the university, he considered dropping out and moving to a kibbutz in Mandatory Palestine, but his intellectual curiosity was reawakened by conversations with the linguist Zellig Harris, whom he first met through common political interests in 1947. Harris introduced Chomsky to the field of theoretical linguistics and he decided to major in the subject. This was how his political engagement gave rise to an intellectual passion for linguistics and a long period of cooperation with his mentor.

For several years he worked with Harris, striving as a loyal pupil to make his system work. During that period, the philosopher Nelson Goodman also became Chomsky's mentor and recommended him for a junior fellowship in the Society of Fellows at Harvard. There Chomsky studied mathematics formal logic and philosophy. He became particularly enthusiastic about working with symbolism. He recalls that period in the 1950s with particular fondness. For several years, nobody exerted pressure on him, forced him to publish or perform other boring academic duties. This freedom and ability to deal with the most unusual issues led to the creation and publication of *Syntactic Structures* in 1957, a book that revolutionized linguistics.

It was as a graduate of Harvard University that he came across a manuscript of Skinner's *Science and Human Behavior*, which was being circulated from hand to hand around the university campus. Chomsky's immediate reaction was one of skepticism. He did not like behaviorism because of the simplifications it engaged in. However, *Verbal Behavior*, in which Skinner sought to demonstrate that language is nothing more than the sum of habits created in the course of individual learning, particularly irritated him and provoked a strong reaction. Chomsky expressed his reluctance toward this concept in an openly hostile review of Skinner's book (Chomsky 1959). He argued that almost every sentence we utter is something we have never heard before, and its use was not reinforced in any way, yet we can create an infinite number of such sentences that our interlocutors can understand. In his opinion, language is creative in its essence and any theory of language should seek to explain this fact.

Chomsky's voice in the discussion over *Verbal Behavior* became one of the foundations on which the edifice of cognitive psychology began to be built, and today it is considered a canonical text in this area. This event was also the moment when Chomsky entered the field of psychology, a discipline he had never dealt with before, and in which he has become a permanent resident. Two of his books, *Cartesian Linguistics* (1966) and *Language and Mind* (1968), were of particular importance in this "invasion". With a freedom natural to himself, Chomsky introduced the concept of the mind into his reflections on language, and thus on a much wider range of human behavior, ignoring the fact that it was at that time a taboo concept for psychologists.

In Chomsky's view, the fact that people have the capacity to master language is a sign of its innate character. Children as young as three years old are able to formulate thousands of sentences that they have never heard before and understand sentences addressed to them by others. This would by no means be possible through learning. Chomsky believes that this ability is based on the innate rules of universal grammar common to all languages. Going further, he believes that our mind must similarly be guided by a specific set of rules of reasoning and perception that we have not yet discovered. He considers it a weakness of psychology that so far we have not managed to achieve what, in his opinion, puts this field at the edge of science. He has even compared it on occasion with pseudoscience.

Professor Chomsky, some time ago you said, "It is quite possible - overwhelmingly probable, one might guess - that we will always learn more about human life and personality from novels than from scientific psychology." This is a very harsh assessment of the field of psychology. Has your opinion changed at all since then?

I don't view it that way. I think it is a tribute to the scope and concerns of psychology—which is, in a way, at the end of the line. If physicists come across a molecule that is too complex, they hand it over to the chemists, and so on down the line. When we reach human beings, it's handed over to psychologists, and in the sciences at least there's nowhere else to go. It's hardly surprising that a great deal is not understood. The same is true, in fact, way down the line to the most basic sciences.

Today, public opinion speaks openly about the crisis in psychology, particularly social psychology. What, in your view, are its causes?

I don't know of the crisis.

So maybe this crisis is not as serious as we psychologists think it is? Regarding research in the field of linguistics, on multiple occasions you have postulated a Galilean style of inquiry. Would you say this is an approach we should also consider following in psychology?

I was actually quoting physicist Stephen Weinberg, who described (and advocated) the "Galilean style, whereby nature is investigated by making abstract models, the investigation of whose properties illuminates the structures of the real world" (Weinberg 1977, p. 175). That's a sensible approach to inquiry generally I think, in psychology as elsewhere. And I think it is pretty much what is adopted in the best work.

Yet many scientists criticize this manner of constructing reality models, in particular dismissing the data that seem to challenge these models. Some people even call this approach "methodological anarchism" (Behme 2016). Could you give us a model example of

such an approach in psychology so that our readers can picture for themselves what it really is?

Just about everything that is done. A published paper in experimental psychology includes experimental results. But every experimentalist knows that constructing the "right" experiment is no simple matter. The first tries give results that make no sense in terms of our (always tentative) theoretical understanding, so the data produced are ignored and the experiments are rethought and redesigned—following the Galilean style. And so it proceeds.

In the 1960s and 1970s, and even into the 1980s, much was said about the need to formulate a general theory to unify the results of research in psychology. You were quite skeptical of the potential for such a theory to be devised, and indeed we have yet to see one; I get the impression that in recent times, psychologists have stopped even discussing the idea, focusing more on the production of sexy research results. What are your views on this subject?

Psychology covers a vast terrain, with problems and results of many different kinds. I don't think we can seriously ask for a "general theory to unify the results of research in biology." And even in physics search for a "unified theory" has proven elusive.

Even if we treat such a general theory unifying research results as a kind of ideal form or utopia, isn't it the case that the aspiration to achieve it results in a wide range of scientific theories that order significant areas of our knowledge? Isn't your transformational grammar such an example in linguistics?

Linguistics is, in my opinion, a central part of human psychology, dealing with core properties of human nature, the most distinctive properties of the species. We can reasonably aspire to a general theory of particular aspects of language, specifically what has sometimes been called the "basic property"—the capacity to construct infinitely many structured expressions of thought that can be externalized in some sensorimotor system, typically sound. That was the goal of "transformational grammar," which in its more modern forms incorporates the essential properties of earlier transformations as special cases of more fundamental rules. This theory should be integrated with theories of acquisition, use, neural representation and evolution of language, and should fall within more general conceptions that include theories of perception, action, growth, etc. And more broadly within biology, and on to the more basic sciences. But it's not clear—to me at least—what a "unified theory of psychology" would be.

In the 1980s, you expressed doubts as to whether psychology could explain such difficult issues as free will. Yet recent decades have brought a range of advances in this area. Has your opinion regarding the potential for grasping the essence of free will undergone any significant revision?

As far as I know, the problem of free will remains where it has been for centuries. In the eighteenth century, it was recognized that much can be explained in terms of determinacy and randomness. We can now say the same in much more sophisticated ways. But there's no sound scientific principle that entails that everything in the universe falls within these bounds. Maybe so, or maybe our science is fundamentally incomplete—or even potential human science, a meaningful concept if human beings are part of the organic world, with capacities that have scope and limits (closely related). There are many sophisticated discussions of free will, moral responsibility, etc., but the basic questions arise in the simplest studies of voluntary motion: my decision to lift my finger to type the next key—or not to. The state of the art in this matter is described by two of the leading researchers on the topic, Emilio Bizzi and Robert Ajemian (2015). As they put it ("fancifully," in their words), we have learned a great deal about the puppet and the strings, but the puppeteer remains mysterious—along with a great deal more in the world, matters well understood and articulated by John Locke and David Hume after Newton's refutation of the "mechanical philosophy" that provided the criterion of intelligibility for the founders of modern science.

One of the few areas of psychology where there have been attempts to create a uniform theoretical framework for the analysis of human behavior is the evolutionary approach, and yet, in spite of this, you have criticized it harshly (Horgan 2000, p. 179). Could you elaborate as to the grounds for your negative approach to this field?

I don't criticize the general approach, and in fact have participated in it myself in work on evolution of language. But I think much of the work fails to observe strictures well described by evolutionary biologist Richard Lewontin in his chapter on evolution of cognition in the MIT *Invitation to Cognitive Science* (Lewontin 1998). He warned of the tendency to construct "Just so stories" about what might have happened through natural selection (incidentally, only one factor in evolution), though without evidence that the story is in fact correct or even any indication of how to find out.

Is it that case that our inability to conduct proper experiments that would provide conclusive data for our speculations about what happened during evolutionary processes means we are doomed to construct "Just so stories"?

In some domains we can hope to develop reasonably plausible theories, using evidence from paleoanthropology, genomics, neuroscience, comparative animal studies. Evolution of language, for example. But in some areas we run up against a wall, and have no ideas about how to proceed, including central aspects of language and thought: for example, the origin of human concepts, which, when investigated carefully, are found to be sharply different from anything known in the non-human world. So we can either concoct stories or adopt Wittgenstein's dictum that "whereof one cannot speak, thereof we must remain silent."

What is your view on the conviction that mental processes are relevant to neurophysiology?

To quote myself 50 years ago, "[I]t is the mentalistic studies that will ultimately be of the greatest value for the investigation of neurophysiological mechanisms, since they alone are concerned with determining abstractly the properties that such mechanisms must exhibit and the functions they must perform" (*Aspects of the Theory of Syntax*, 1965). The position is somewhat similar to David Marr's famous discussion of levels of explanation in biology (Marr 1982), and the primacy of the computational level. And in a sense, it's the "Galilean style." There have been some exciting developments in the area of my own special interests in finding neurophysiological evidence bearing on fundamental properties of language.

Sticking to the subject of neurophysiology, I would like to ask you about one particular discovery—mirror neurons, which have become a particularly fashionable subject in recent years, and in particular

I want to ask about the role ascribed to them in the creation of language. I know that advocates of your conceptions are quite critical of such ideas, but I am curious as to your personal opinion on the subject.

These are very important discoveries, but I haven't seen any convincing argument that they bear on the evolution, acquisition, or use of language in any important way.

In discussing mirror neurons and their role in the creation of language, we are reaching the borders of science. Many pseudo-scientific conceptions invoke well-known scientists and complex ideas, which they frequently do not understand, such as quantum physics. This is indefensible, but I believe we need to speak the truth about this openly. Your name and your conception of transformational grammar have appeared in the company of the theories of Gregory Bateson and Alfred Korzybski, as well as ideas and techniques from Carlos Castaneda, as the foundations of neurolinguistic programming. A search on Google using the terms "Chomsky" and "neurolinguistic programming" returns over 300 thousand results. Unfortunately, I am unable to find in any of your public statements anything about your position in this matter. Would you take this opportunity to address it?

Not really. The last time I looked at Korzybski's work was when I was an undergraduate 70+ years ago. I wrote some papers arguing that it was valueless. I know little about Castaneda. Neurolinguistic programming purports to adopt ideas from generative grammar, but I know no more about it than that.

Let's take a closer look at your achievements. Which of your discoveries do you consider the most important in the development of science, and why?

At a very general level, I think it was important to formulate what seems to me the "basic property" of the faculty of language, described above, as a biological trait, a property of an individual, which can be studied by the normal methods of science. If this sounds close to truism, it should. And in fact it revitalizes in new ways a rich and long forgotten tradition, though it was a departure from then-prevailing behaviorist and structuralist currents and from a great deal of current work as well.

Adopting this perspective, there are many tasks, and I think there has been quite considerable progress in addressing them, which I cannot review here. I think, in fact, that it is now possible to seriously contemplate, maybe even to achieve, the long-sought objective of accounting for some fundamental properties of language in ways that meet the twin conditions of learnability and evolvability.

Is there any particular type of criticism aimed at you that you feel is particularly serious and justified?

The criticisms I know of I've either incorporated in my own work or sought to respond to, how successfully is for others to judge.

The power of science is found in the questions it asks and the answers it provides. What in your opinion great questions has psychology managed to answer so far?

That question could be answered seriously only by someone who has mastered the vast range of topics encompassed by research in psychology. Perhaps there is such a person. It certainly isn't me.

What about significant questions still waiting for answers?

Same problem. I'm sure there are a great many of enormous significance. In the area of my own specific concerns, one critical question has been raised and studied by cognitive neuroscientist C.R. Gallistel with A.P. King (2009) in very important work: what is the neurophysiological basis for the remarkable computational capacities of insects and other organisms, and in a very special and unique sense, the human faculty of language? He and others have given strong arguments that the answers are not to be found in neural nets, more likely in internal properties of neurons, perhaps down to the cellular level.

Do you think search for answers to these questions can be aided by the intensive growth of AI that we are witnesses to?

Perhaps some day, but I think there are serious limits in the interesting work in AI that has so far taken place, matters I've discussed elsewhere (Chomsky 2012).

On multiple occasions you have been highly critical of the modern education system, while at the same time emphasizing that you were lucky to have begun your career in conditions far more conducive to intellectual development. Are there presently any alternatives outside universities for the development of young people?

I happen to have had a highly idiosyncratic educational experience. I benefited greatly from the enormous intellectual wealth offered by great universities, but without fulfilling professional requirements in any established field, including linguistics. I don't know of any comparable resource.

Who would you point to young people just getting their careers started as an example to follow?

I'm not a great believer in models. I think each of us has to find our own way, and a proper educational system should facilitate that process. The great linguist and humanist Wilhelm von Humboldt, founder of the modern research university, once described ideal education as a process of laying out a string along which students pursue their own course. There should, that is, be some degree of structure, ensuring that students will benefit from the wealth of human achievement, but with ample scope for creative initiative and exploration by independent minds.

Is this even possible in a system about which you said "The whole educational and professional training system is a very elaborate filter, which just weeds out people who are too independent, and who think for themselves, and who don't know how to be submissive, and so on – because they're dysfunctional to the institutions"?

That's certain a common, often prevailing feature of educational systems—something pointed out by Orwell incidentally in his unpublished introduction to *Animal Farm.* But higher educational institutions at least, and often others, usually provide opportunities to escape and explore the riches of human cultural achievement, and to add to them.

Professor Chomsky, beginning in your childhood and continuing throughout your whole life you have been a harsh critic of political elites and the existing social order, pointing to their dark sides. Not long ago, Steven Pinker used statistics to present a vision of the world that we live in today as the best humanity has ever experiences. Some go so far as to call it the "grand apology for capitalism". What are your views on the subject?

No one has been more of a cheer leader for capitalism than Karl Marx—who, however, also recognized and explored its much darker side, which does not show up in statistics. That includes the "metabolic rift," the inherent tendency of capitalism toward ecological catastrophe, a critical issue right now. Those who are not hiding under a rock understand that the threat of global warming to organized human life is severe and imminent. Among those who understand it very well are the managers of the energy corporations, who are racing to increase fossil fuel production, and the CEOs of the great banks, who are pouring resources into the most dangerous of these, like Canadian tar sands. We can also add the management of Royal Dutch Shell, now exulting in the construction of an immense plant to produce non-biodegradable plastics, in full awareness of what this means for the oceans. And the Trump administration, whose own environmental assessments anticipate that by the end of the century temperatures will rise to 4 °C above pre-industrial levels, about twice as high as what the scientific community regards as the limit for organized human life to persist, and are therefore racing to maximizing the destruction. All following impeccable capitalist logic. A real "grand apology for capitalism." We can add a lot more to the grand apology. To mention just one of myriad examples, there is real need to develop modes of carbon capture, but market signals instruct venture capitalists to ignore such long-term projects with dubious profits in favor of new apps for I-phones, so the research languishes, contributing to the doom of organized society under capitalist logic. And I should perhaps add that little homilies about how science and technology always rise to the challenge don't help, to put it mildly.

Turning to the material improvement, there is serious work, notably Robert Gordon's study on the rise and fall of American growth, which generalizes beyond (Gordon 2016). He shows that there was virtually no economic growth for millennia until 1770, slow growth for another century, and then a "special century" until 1970, dependent largely on specific inventions ranging from indoor plumbing to electrical grids, the automotive industry, etc. Since the 1970s the picture is much more mixed, including the nature of inventions. In the US there has been GDP growth, though slower than before. But it has gone into very few hands. Real wages for non-supervisory workers has declined, and mortality has even increased in recent years. Inequality has increased to such grotesque levels that half the population now has negative net worth, and the tendencies are increasing. Gross statistics miss all of this.

And much more. Poverty worldwide has reduced, largely because of China, hardly a model of capitalism (which should not be confused with markets). One of the most remarkable success stories is Russia, in 1917 a very poor peasant society, virtually a neo-colony of the West. Despite the devastating war, the immediate Western invasion, and the hideous destruction of World War II and the harsh and brutal totalitarian social structure, it became a major industrial society, its satellites in tow. Hardly a triumph of capitalism. Furthermore, serious inquiry would tell us not just what is but what could be with decent policies, given the resources and options—particularly in the richest and most powerful country in world history, with incomparable advantages.

Without continuing, unanalyzed statistics tell us very little about the world.

Would you agree to answer one of the 30 questions that my readers would like to ask the most eminent psychologists in the world?

Not sure what you mean. I don't know what the questions are, and doubt very much that I'd be competent to answer them.

One of them is: The idea of psychology, which we are the most ashamed of? Would you like to answer it?

Don't know how. Psychology has exciting achievements. It's often gotten lost in byways, which is not a matter of shame but rather something to learn from. One part of psychology, the study of the human mind, is one of the most difficult topics of science, and we don't need the Delphic Oracle to instruct us about its significance for human thought and life.

Selected Readings

Berwick, R., & Chomsky, A. N. (2016). *Why only us: Language and evolution.* Cambridge, MA: MIT Press.

Chomsky, A. N. (1964). *Current issues in linguistic theory*. The Hague: Mouton.

Chomsky, A. N. (1975a). *Reflections on language*. New York, NY: Pantheon Books.

Chomsky, A. N. (1975b). *The logical structure of linguistic theory*. New York, NY: Plenum Press.

Chomsky, A. N. (1980). *Language and learning: The debate between Jean Piaget and Noam Chomsky* (M. Piattelli-Palmarini, Ed.). Cambridge: Harvard University Press.

Chomsky, A. N. (1982). *Language and the study of mind*. Tokyo: Sansyusya Publishing.

Chomsky, A. N. (1984a). *Modular approaches to the study of the mind*. San Diego, CA: California State University Press.

Chomsky, A. N. (1984b). *Modular approaches to the study of the mind*. San Diego, CA: California State University Press.

Chomsky, A. N. (1986). *Knowledge of language: Its nature, origin, and use*. New York, NY: Praeger.

Chomsky, A. N. (1987). *Language in a psychological setting*. Tokyo: Sophia University.

Chomsky, A. N. (1988). *Language and problems of knowledge: The Managua lectures*. Cambridge, MA: M.I.T. Press.

Chomsky, A. N. (1995). *The minimalist program*. Cambridge, MA: MIT Press.

Chomsky, A. N. (2000). *The architecture of language* (N. Mukherji, B. N. Patnaik, & Rama Kant Agnihotri, Eds.)Rama Kant Agnihotri). Oxford: Oxford University Press.

Chomsky, A. N. (2006). *The Chomsky-Foucault debate: On human nature*. (with Michel Foucault). New York, NY: The New Press, distributed by W. W. Norton.

Chomsky, A. N. (2009). *Of minds and language: A dialogue with Noam Chomsky in the Basque Country* (M. Piattelli-Palmarini, J. Uriagereka, & P. Salaburu, Eds.). Oxford: Oxford University Press.

Chomsky, A. N. (2012). *The science of language (Interviews with James McGilvray)*. Cambridge: Cambridge University Press.

Chomsky, A. N. (2015). *What kind of creatures are we?* New York: Columbia University Press.

References

Behme, C. (2016). How Galilean is the 'Galilean method'? *History and Philosophy of the Language Sciences*. Retrieved from https://hiphilangsci.net/2016/04/02/how-galilean-is-the-galilean-method.

Bizzi, E., & Ajemian, R. (2015). A hard scientific quest: Understanding voluntary movements. *Daedalus, 144*(1), 83–95.
Chomsky, A. N. (1957). *Syntactic structures*. London: Mouton.
Chomsky, A. N. (1959). A review of Skinner's verbal behavior. *Language, 35*(1), 26–58.
Chomsky, A. N. (1965). *Aspects of the theory of syntax*. Cambridge, MA: MIT Press.
Chomsky, A. N. (1966). *Cartesian linguistics: A chapter in the history of rationalist thought*. New York, NY: Harper and Row.
Chomsky, A. N. (1968). *Language and mind*. New York, NY: Harcourt, Brace & World.
Chomsky, A. N. (2012, November 1). Noam Chomsky on where artificial intelligence went wrong. (Interviewed by Y. Katz). *The Atlantic*. Retrieved from https://www.theatlantic.com/technology/archive/2012/11/noam-chomsky-on-where-artificial-intelligence-went-wrong/261637/.
Gallistel, C. R., & King, A. P. (2009). *Memory and the computational brain: Why cognitive science will transform neuroscience*. New York, NY: Wiley-Blackwell.
Gordon, R. J. (2016). *The rise and fall of American growth: The U.S. standard of living since the civil war*. Princeton, NJ: Princeton University Press.
Horgan, J. (2000). *The undiscovered mind: How the brain defies explanation*. London: Phoenix.
Lewontin, R. C. (1998). The evolution of cognition: Questions we will never answer. In D. Scarborough & S. Sternberg (Eds.), *An invitation to cognitive science. Methods, models, and conceptual issues: An invitation to cognitive science* (Vol. 4, pp. 106–132). Cambridge, MA: MIT Press.
Marr, D. (1982), *Vision: A computational approach*. San Francisco, CA: Freeman & Co. MIT Press.
Skinner, B. F. (1953). *Science and human behavior*. Oxford: Macmillan.
Skinner, B. F. (1957). *Verbal behavior*. New York, NY: Appleton-Century-Crofts.
Weinberg, S. (1977). The forces of nature. *American Scientist, 65*, 171–176.

11

Roy F. Baumeister: Self-Esteem, Self-Control and the Power of Will

Do better each time.

Roy F. Baumeister

T. Witkowski, *Shaping Psychology*,
https://doi.org/10.1007/978-3-030-50003-0_11

Rarely do we appreciate the role of chance in history. Even less often do we uncover seemingly insignificant facts that could have initiated a chain of events leading to far-reaching historical changes and events perceptible by all. There is every possibility that the history of social psychology would be entirely different today if not for the trivial and insignificant fact that in the 1970s, Standard Oil's staff psychologists were paid more than some of its managers. It was precisely this fact, despite initial reluctance, that ultimately convinced a certain manager at Standard Oil to allow his son to change his major at university from mathematics to psychology (Storr 2014). Perhaps if it were not for this simple fact, social psychology would have been deprived of one of its most creative and prolific researchers, an individual responsible for changing the face of the field multiple times and who, in 2013, was given the highest award of the Association for Psychological Science, the William James Fellow award, in recognition of his lifetime achievements.

As a student, Roy Baumeister sympathized with the notion that low self-esteem was at the root of most social problems. This way of thinking was largely propagated by Nathaniel Branden's blockbuster book *The Psychology of Self-Esteem*, published in 1969. Baumeister read it, but apart from convincing anecdotes, it lacked science. To fill in the gaps, he started studying the differences between people who think highly of themselves and those with low self-esteem. His interest was partly pragmatic: Self-esteem was a hot topic and a great subject for a young academic beginning his career. His undergraduate thesis explored how people respond to public challenges to their self-esteem. Gradually, however, he felt a growing sense of doubt, which took on urgency in the mid-1980s. On the one hand, it was fuelled by his own research results, which showed that people with low self-esteem are in fact not aggressive and prone to violence, but rather shy and submissive toward others. On the other hand, it turned out that there was very little hard data in support of many popular notions.

In 1996, Baumeister co-authored a review of the literature that concluded that it was "threatened egotism" that in fact led to aggression. Evil, the authors suggested, was often accompanied by high self-esteem. "Dangerous people, from playground bullies to warmongering dictators, consist mostly of those who have highly favorable views of themselves," they wrote (Baumeister et al. 1996). Such statements turned widely held convictions about the role of self-esteem on their head. In 1999, the Association for Psychological Science (APS) asked Baumeister to lead a team to review the literature in its entirety to see, finally, what effect self-esteem had on behaviors such as happiness, health, and interpersonal success. Their study, published in May 2003, came as a shock to many researchers and practitioners. It concluded that efforts to boost self-esteem failed to improve school performance. Nor did self-esteem help in the successful performance of various tasks. It didn't make people more likeable in the long term, nor increase the quality or duration of their relationships. It didn't prevent children from smoking, taking drugs, or engaging in "early sex." Self-esteem did, however, enhance mood and seems to support initiative (Baumeister et al. 2003).

Apart from self-esteem, Baumeister inquired about the reasons for self-defeating behavior. He concluded that there is no self-defeating urge (as some have thought). Rather, self-defeating behavior is either a result of trade-offs (enjoying drugs now at the expense of the future), backfiring strategies (eating a snack to reduce stress only to feel more stressed), or a psychological strategy to escape the self—where various self-defeating strategies are rather directed to relieve the burden of selfhood.

In 1995, together with Mark Leary, Baumeister co-authored the need-to-belong theory, which seeks to show that humans have a natural need to belong with others. Baumeister and Leary suggest that human beings naturally push to form relationships. The diverse topics they explore in their research include irrationality and self-defeating behavior, the influence of culture and nature on human sexuality, erotic plasticity, decision making, and the meaning of life (Baumeister and Leary 1995).

Perhaps, however, his research on self-regulation and willpower has brought him the most fame and recognition. By taking up these extremely demanding and important issues, he placed himself in the front row of a small group of researchers who restored the once-central issue of strong will to the mainstream of psychology after half a century of neglect. It is hard to overestimate the importance of these studies, as self-regulation processes seem to be crucial in perhaps all areas of our lives. Our success in school depends on them, as does the possibility of achieving our goals, establishing positive habits and fighting against the bad ones. Baumeister has authored a volume on self-regulation, *Losing Control*, has edited three editions of *Handbook of Self-Regulation* (Baumeister and Vohs 2004) and has devoted numerous experiments and journal papers to the issue. He also describes this research in a book, *Willpower*, authored with former *New York Times* journalist John Tierney.

Recently, he has been absorbed by the search for an answer to questions about the importance of free will and consciousness. He approaches these topics from the perspective of evolutionary psychology. In his view, both free will and consciousness are products of evolution taking place within the context of culture. They are advanced forms of action control that allow humans to act in pro-social ways toward their enlightened self-interest when acting in these ways would otherwise be in conflict with the fulfillment of evolutionarily older drives or instincts.

Professor Baumeister, at the beginning of your scholarly path you were adherent of an ideology that identified the causes of most individual and social problems as low self-esteem, while its enhancement was supposed to be a panacea for all sorts of illnesses. Today you are referred to as the man who destroyed America's ego. Was your conversion a sudden one, or did it result from a gradual evolution of your views?

It was gradual. I'm not sure how much I believed at the beginning that self-esteem was a cure for everything. I thought it was a positive thing. And then when other people started saying "look at all the good things correlated with high self-esteem," I started to support them. The research I did was not so much aimed at curing problems or making people better, but rather at understanding differences in social strategies. Solving problems, that was really other people's work, and I was glad to hear that because it would've made my own work more important.

I remember the first hint was from a sociologist at a meeting, who said "how come self-esteem doesn't seem to ever do anything?" I was surprised, and I responded "no, there are meaningful differences with self-esteem," and I shared some of my findings with him. But it stuck in my mind that he was somebody who had looked at other findings from the big surveys that sociologists do and noticed that self-esteem didn't seem to predict much of importance at the societal level. So, I started paying attention to that, and gradually I came to see more and more.

But with self-esteem, we were expecting too much from it, that people with high self-esteem were not really better.

The research on aggression was a turning point for me. I was doing research for a book on evil and I'd heard that low self-esteem was associated with greater violence. By then, I knew enough about people with low self-esteem that I was skeptical, because in our work people with low self-esteem lack self-confidence, they avoid risks, they go along with what others say. None of that goes with being highly aggressive, which is a risky dominating strategy, but I believed it because the research was there. And when I started looking at the literature, there was no evidence that low self-esteem was associated with violence. In fact, quite the opposite—the aggressive people were those who thought they were better than everyone else, and they got mad when someone questioned them. That was a pivotal change, a turning point at which I started becoming more skeptical of the benefits of high self-esteem in general.

I remember that California formed a task force about self-esteem and they published an edited book. I read it, and the evidence for the benefits of high self-esteem was pretty thin, even though it was produced by people who were really trying to promote it and advocate it. That also made me skeptical. Next, the APS appointed a number of us to re-evaluate the benefits of high self-esteem. I don't know if I was a skeptic, but I said "let's go and look at all the evidence again." We assembled a team that included some people who were very positive about self-esteem, some who were very skeptical and some in-between. We converged around the conclusion that there were a few benefits, but nothing spectacular.

In your public statements you criticize the so-called authentic self, true self or self-actualization, which is commonly employed by pop psychology and the psychotherapeutic industry. Could you elaborate a little more on your reservations toward these concepts?

There's a recent article (Jongman-Sereno and Leary 2018) that does a better job than I have of documenting the problems. First of all, the belief that you have a true self that's different from the way you act is questionable. Second, since everything you do reflect you. While you might sometimes want to distance yourself and say "I didn't mean that," or "I don't wish to be known as the sort of person who does that," the fact remains that you really did those things. They are expressions of yourself.

The authentic self is a kind of mythical creature. A special journal issue is coming out on this, and many of the papers complain that when people talk about their true self, it might be different from how they view their actual self—but somehow the true self is always good. People seem to think their true self has only positive traits. That's already questionable, as people who believe they are endowed with "good" true selves go on to behave badly. What I see happening is a convergence on the truth of a sort of ideal or myth that you strive to become, so when you act in a way that's consistent with that you say "yes, that's the true me," and when you act inconsistently with it then you are at a distance from yourself—"ah, no, I don't want to be viewed that way." But they're both the real you.

Fast-forward to skepticism about self-actualization. Again, if it's the idea that you're born with some destiny that must be realized, I'm skeptical of that. However, in terms of self-improvement and so on, there is a difference between fulfillment and unfulfillment. I'm not sure that there is only one "you"—you might have a fulfilling life as a father who stays home and raises his children, or a farmer who manages his crops well and produces food, or the same person could find fulfillment as a musician. At the same time, you could fail to find fulfillment doing any of those things as well. What I'm skeptical of is the notion that we're born or endowed with one particular version of "self" that is different from the way we really are, but you need to become that person. That doesn't fit with most of what we know about scientific psychology.

Are there really no psychologists or therapists with the capability to decide in an arbitrary manner which of our emotions and thoughts are part of our true self and which are not?

That's a good question. There's an article in which two scholars coin the term "inevitable authenticity" (Jongman-Sereno and Leary 2018). Everything you do is a product of you. This idea of true self comes more from the experience of false self-behavior, when you feel that what you did was "not you," and sometimes you're pressured to go along with the group and do things that you regret. So there is some legitimacy to that point; again, you really did do those things, you did decide to go along. But we have moments of weakness, or perhaps we didn't understand the full implications. So I'm willing to give some credence to false self-behavior.

I think there's a logical fallacy—people think that because there are false versions of self, there must be a single true one. I think there might be multiple true ones, some of which are truer than the false ones, but the fact of having many false selves does not imply that there is one true self. Again, what people often mean by "false self" is that they want to distance themselves from it. They wish they hadn't done something that they did. We all end up wishing we hadn't said or done certain things. Maybe there are people who look back at their lives and say "every choice I made was perfect," but I've certainly never met anyone like that. Most of us think "my life's gone pretty well, but there are things I regret and that I wish I had done differently. I want to distance myself from those things." The creation of the self is a continuous work in progress.

I know some therapists who are convinced that they can clearly distinguish authentic self from not authentic self…

Good for them. If they could publish research, that would be a valuable contribution for everyone else to know. Another problem with the "true self" in the authenticity literature is that it's all self-report. We ask people "do you feel authentic?" and yet we know people deceive themselves, they engage in wishful thinking. How can we rely on self-report as the ultimate criterion for whether something is the true or the false self, given that people are wrong?

Self-reports is a big issue in research methodology and we'll return to it. But I'd also like to ask you about the importance of self-esteem and its defense mechanisms. This is a very prominent research trend in psychology, crowned with such conceptions as terror management theory, according to which the need to maintain and boost self-esteem is the primary motivation underlying most of our behaviors. What is your attitude toward this and similar notions?

I think that's wrong at multiple levels. First of all, terror management thinks that fear of death is a primary motivation, and that the desire for self-esteem is a derivative of that. I don't really understand this conception because the higher your self-esteem, the worse death is. If you have low self-esteem, you think have no value. The fact that you're going to die doesn't really matter that much. But if you are such a marvelous person, then your death is a huge tragedy.

Here in Australia, the need for positive social identity has been put at the center of some theories of psychology, to explain why people join groups and organizations. But I don't see why that would evolve. To me, the desire for a positive view of self must be derivative of other motives that are more closely tied to survival and reproduction. The benefits of high self-esteem are very limited. What are those benefits? We looked at great length and found two: initiative, that is, you have more confidence in yourself, and it feels good. But that's not very much, and the fact that it feels good itself has to be based on something else. Nature didn't just link something to good feelings for no reason. Good feelings evolved mostly because they're associated with doing things that help you survive and reproduce. Getting food when you're hungry feels good, just like getting water and having great sex, seeing your children do well—all these things are associated with positive feelings. Self-esteem contributes very little to any of those.

I've worked with Mark Leary on his sociometer theory. A fairly good explanation of the desire for self-esteem is that it's your inner meter for keeping track of how you're doing socially—will other people accept you and like you, will they hire you, will they marry you, will they sleep with you, will they take you for a friend, and so on. Those things are important for survival and reproduction, and your self-esteem is your inner measure of how you're doing on those. But just thinking you're good without being good is not much help, although it is seductive. It's a bit like taking a drug to feel like a successful person rather than actually achieving something.

There is also another group of European scholars who said that what counts is not just acceptance by others, but also climbing the social hierarchy. That's also true. I think they're quite right that we evolved in primate groups whose higher-ranking members had better chances for both survival and reproduction. So, yes, we humans evolved to strive for success both in terms of being accepted and in terms of getting to the top of the hierarchy; self-esteem is our assessment of how we're doing and where we're likely to end up in that regard. So that's important, and we need to have some internal measure of it.

Defense mechanisms are a sort of flaw in the system. Again, it's a bit like taking drugs that give you the great feeling of achieving something tremendous even though you didn't really achieve anything. But the basic mechanism is to get us to do things that advance our standing in the social world that, in turn, improve our prospects for survival and reproduction. Motivation starts with the needs instilled by biology. Creatures who survived and reproduced spread their genes, while others do not, and so psychology's traits are based on those motivations to do the things beneficial to reproduction.

A counterweight to people who continue to profess the cult of self-esteem and often earn a good deal of money or make careers in science are those who deny the existence of the self at all. As far as I know, you also disagree with them. Could you elaborate as to why?

Wanting to see what the case was against the self, I just read Bruce Hood's book *The Self Illusion* (2012). He didn't make a very good case by saying the self is not quite what it seems. I'd say to some of these people "that's all good and well, so the next time you have to take an international flight, don't take your passport and just explain to everybody that selves are illusions." It doesn't matter who's checking boarding passes, just say that to them and see how that goes. It's ridiculous. Our whole society functions on the basis of those selves, we have possessions and relationships, and people are not freely interchangeable. Even if you don't know who you're sleeping with, your wife knows who you're sleeping with, and if you're sleeping with somebody else you can try and say to her "oh, well, there's no such thing as the self." See how that goes.

Suicide can be quite good evidence of the reality of self. Recalling your thoughts on the subject from nearly 30 years ago (Baumeister 1990), is suicide a form of escape from self?

Yes, I think that theory remains basically correct, but it's not a complete explanation. There are multiple reasons why people commit suicide. I know of another major theory by my friend Thomas Joiner (2005) in which he's made a number of excellent points. So again, mine is not a complete explanation. However, it starts with a spoiled identity and the feeling that I can't stand being who I am. Suicidal thoughts are a way of getting rid of that unpleasant self-awareness. Suicidal people have usually experienced something bad that they feel responsible for, and which reflects badly on the self. So they have an intense, negative awareness of themselves, and like with other kinds of escape from self-awareness they resort to something to blot out that awareness. Suicide, even an unsuccessful attempt, usually takes you out of your life, and during the attempt, it focuses the mind very narrowly on the here and now. It takes you away from the more problematic aspects of self. So, yes, people do all sorts of things to escape from awareness of themselves, especially when that awareness is bad.

Your first iconoclastic text concluding that the perceived importance of self-esteem is overrated was probably the chapter titled "The Self" in *The Handbook of Social Psychology*, published in 1998. It was followed by more texts in the same vein. After more than a quarter-century, would you say that you succeeded in "destroying America's ego" and that self-esteem is today merely a historical curiosity?

No. I and other people have raised some questions. I was surprised to see journalists using that title, but then again, a headline is a way to catch attention. So it was amusing, but it's certainly not how I think of myself. I don't remember what my first published work questioning self-esteem was; I think the *Psychology Review* article "The Dark Side of High Self-Esteem" might have been the first one (Baumeister et al. 1996). Anyway, in terms of where we are today, the questions that I and many others raised have put something of a damper on it, but schools still talk about boosting kids' self-esteem. They still think it's important, but I think that's a misguided effort. I think it's actually destructive in some ways, as I've come to think that self-control is much more valuable for solving problems.

Good self-control is good for both the individual and society. High self-esteem, if it's good at all, is good for the individual. If you have someone with an inflated view of self, it might be nice for them, but the costs are borne by the people around them. If you've ever worked with somebody who had a big ego and thought they were God's gift to the world, you know what I'm talking about. It's difficult to get along with those people. However, self-control is good for everyone. And so my advice to schools, parents, coaches, teachers and everyone else is "forget about self-esteem and concentrate on self-control."

The cost is in practice when parents get so worried about their children's self-esteem than they don't want to criticize, and so they don't do the things needed to build self-control, which is to set rules and insist that the child live up to them. Instead, they just tell their child "Dude, you're perfect. You're wonderful. You're doing everything great." And that's a harmful message.

Why, in your view, do some psychological conceptions enjoy such tremendous popularity despite research results that consistently negate them?

The self-esteem movement was very enjoyable to participate in as it's fun to think about how great you are, sit around in a group where everyone tells everyone else how wonderful they are and "this is what I like about you." I sometimes joke that American schools have always been looking for ways to get students to learn better without actually having to do more homework. So this idea came out: "Well, we could just tell kids they're good at math, and if they believe they're good at math then they'll be better at math. They won't have to do the math homework." But I don't think it works.

You really have to do the math homework, but certainly spending an hour taking turns about saying how wonderful we are and appeasing each other as opposed to spending an hour doing math is going to be much more fun for everyone, including the teacher. So I can understand why it's popular. I just don't think it really does any good.

Jeremy Freese said that such conceptions are more vampirical than empirical because they are unable to be killed by mere evidence. But to conclude the conversation on self, self-esteem and self-defense mechanisms, I would like to ask you about the attitude of the scientific community to your views. After all, they contradict many widely held opinions. Is it difficult to be a scientific dissident?

Yes, it's certainly difficult to be a dissident of any sort and yes, the creativity of science comes from people questioning what's there and coming up with alternative theories. Many of these alternative theories turn out to be wrong, but we need to have them. It is difficult to be in that position, especially when other people like the dominant theories, or people made their careers doing things a certain way and you say "no, you need to do things differently."

And also, it's fair to expect a heavier burden of proof being put on the dissident. If you're questioning something that everyone believes, you need more data than if you're going along with the mainstream because often what everyone believes is correct, or at least it's the best that we can do now. We need people to question and challenge. But it is fair to say "you need stronger evidence if you're going to go against what everyone has been thinking."

In your research, you seek answers to great fundamental questions that have been the subject of philosophical considerations for centuries, and which went neglected by psychology for decades. I am thinking here mainly of willpower or free will. In my eyes, this is an expression of great courage, because it is safer and easier to conduct research on academic intelligence, which will certainly give publishable results, than to risk failure trying to answer such difficult questions. However, apart from courage, there is also the practical aspect of conducting such research. Many of my interviewees admit that they would never receive grants for doing research on what they are really interested in, because the preferences of the institutions responsible for making decisions about funding were completely different. When applying for funds, they often had to compromise. How is it in your case? Is it easy to obtain funds for research in which such difficult questions are asked?

It's tough. The work on self-control and self-regulation did do fairly well with funding, but in general that's correct. To go against established views and even to switch to new topics, which I've tried to do—that is difficult to get funding for. I briefly was going to be a philosophy major. I went to university to do mathematics first, and then I wanted to do philosophy. When I went into psychology, as a sort of compromise, I thought "well, what if maybe we should do that?" I did my laboratory work, the basic work you need to establish your credibility as a scientist. Then, I was able to write a series of books addressing various big philosophical questions. With social science, data is one of the things that got me into psychology in the first place. I had been reading moral philosophy. And when I read Freud, instead of trying to analyze the concepts of what is right and what is wrong and how we know if something is truly virtuous, Freud's approach was: "Let's look at how people actually learn concepts of good and bad, right and wrong." I thought, "that's a great way to approach and get data. You can address the same problems with a fresh look." I've tried to do that.

My first book was on identity, a book on human nature. I do have thoughts of doing a free will book at some point. *Willpower* doesn't really address the grander issues of free will, consciousness. So, I am working around or through these notions in a way, but I've also had in mind that I do these big philosophical projects just on the side. I do them as literature reviews, because they require big thoughts, and you can't do an experiment to settle the meaning of life. So I've kept my laboratory, working on one or two problems over time. For 20 years, I had two main themes in my laboratory: One was work on self-control, self-regulation, ego depletion, and the other was rejection and belongingness. I managed to get grants for one or the other, going back and forth, and that was enough to fund the laboratory.

But my work didn't actually need grants. It was nice to have them, but I was able to do psychology just with the subject pool in the laboratory facilities. And as long as there was support for Ph.D. students, then I didn't need grants. I've had grants more or less continuously since 1992, the last one ended just about a year ago, so I don't have a grant right now. But for a long time, one seemed to come along after another. I thought that each one might be the last, and I moved to Florida State partly because they gave me a good research budget. That meant if I didn't have a grant I could still pay for research expenses that came up.

What strategy for raising funds for research should young researchers adopt today? Formulate their research questions directly regardless of the attitude of the institutions granting funds, or rather try to adapt to fashionable research trends?

This is a challenge. I want to encourage young people to do what they want and to do the science based on their own best ideas, but to the extent you need funds to do that you have to adapt. And I think it is legitimate for the government or the funding agencies to say "these are our priorities and if you don't want to do research on this, we're not going to give you a grant." Many people find ways of compromising. So you say, "my research has a little bit of relevance to cancer, so I could get a grant from the Cancer Institute and do some work on cancer. But for part of that I can also do what I want, something more creative and in line with my own ambitions." I think that's legitimate, and that's not taking advantage of the system—it's working with the system so that we supply the kind of research that the granting agency wants while also getting our own work done.

In some fields like social psychology, it's not absolutely necessary to have grants to do some research. But if you're in a field that simply needs grants, you have to go along with what's available. There are many sources of grants, not just the government and private foundations. I've had grants from a range of groups, including those. It's something you have to be aware of because their priorities change. In America, the National Institute of Mental Health was a main funder of social psychology for decades, and then at some point they just stopped and said "we're not going to do that anymore." A lot of people who were working with that system suddenly were left out with no grant money. The government doesn't care about social psychology. They're trying to think about how we can make society better. So they thought, "we need to spend money on mental health and maybe social psychology can help with that." Then at some point they looked it over and said "all this money we've spent on social psychology has not really helped the mental health of our country, so let's not spend on that anymore."

So, I told some of my young people "maybe look at physical health. Maybe there's something we can do there, certainly problems like addiction have a psychological aspect." Addiction is a good subject that's not going away, and there are a lot of things to learn about it tied into self-control and self-concept, and many other things you can study in connection. There's also things like arthritis or diabetes, many things have behavioral aspects. So you have to find a way to link what you want to study with what the government or some private foundation sees a need for in society and wishes to put money into.

Changing the subject: today many philosophers and scientists, including some of my interlocutors in this volume, say openly and directly that free will is an illusion. You yourself have devoted a lot of energy to research on free will, and, as far as I know, you're of the opposite opinion. What are your arguments for free will?

I was interested in it for a while. I got the idea because I was interested in self-control and rational decision making, two important forms of free will. It turns out that the people who say there's no free will—and I have debated with several of them—more or less agree with me about the psychological processes that are involved. It's just a question of whether these should be called free will or not. I think it's really not much of an argument worth having anymore, but more a semantics dispute. If the people who are against free will tend to think that free will means that this is an exemption from causality, or that we have a soul that causes behavior independent of physical and psychological causes, I don't believe any of those things. So I agree with them saying that there's no free will in that sense.

The people who believe in free will say "it's just the ability to act differently in situations," and I very much believe in that. In fact, I'm not sure you could do psychology without believing that people have the capability to act differently in the same situation, that people make choices and that there are real multiple possibilities. Moral judgment essentially is "should I have acted differently in that situation? Did I do the wrong thing? I should have done something else." It's based on the assumption that you could have done something else. Moral judgments seem predicated on this basic kind of free will.

The questionable aspect to me is not so much the freedom part, but the will part. Is there such a thing as a will? There is an agent. Humans have evolved ways of controlling our behavior and deciding what to do that are qualitatively different from other animals. We can project into the future and we can invoke moral principles, we can do economic calculations. We can have long-term commitments and considerations and then change how we act on the basis of that kind of thinking. So that's the reality. Whether you call that free will or not is just a semantic argument. So I don't debate free will much anymore. This is funny, I'm giving a lecture on it tomorrow and that's what I'm going to say. But I have talked at some length about these processes and how they evolved, and why they are different from what other creatures have. It's not clear that other animals really have a full-blown moral sense the way humans do, nor the ability to represent the future in terms of multiple possibilities. Again, that seems limited to human beings. That's the psychological reality, the exciting puzzle. Whether we call it free will or not is kind of a sideshow.

Your search for answers to basic questions about the meaning of human existence was directed by your childhood inclinations toward the Bible, and as you've mentioned, as a student you become fascinated by the writings of Freud. What is your attitude to these early sources of inspiration today? Do they play any role in your search for answers?

I'm not sure the Bible was a big inspiration. I was brought up in a moderately Christian family, but my parents weren't very strict about it. They sort of believed, but they didn't go to church very much. They made me go to Sunday school, which I found terribly boring. At some point I said, "can I just stay home and read the Bible instead of going to Sunday school?" Being Protestant, they thought that sounded reasonable. So I read the whole Bible cover to cover. But I'm not sure it was a huge inspiration to me. I did wonder about the teachings, and I was questioning and skeptical from an early age.

When I was studying philosophy, I was also interested in religion, and I wondered whether all the religions of the world could just be products of the human mind, or of other psychological needs that made up all these religions. I think the answer is yes. That doesn't mean that they're all false. They contradict each other, so most of them are false, but there could be one "true" religion. But even then, the reason we believe is because of our own psychological inclinations and propensities. So, to me, religion is not a big inspiring source of the meaning of life.

Freud had a big influence on my thinking. I admired the discipline and the wide-ranging scholarship and the careful reasoning. Freud is kind of fading away from psychology, and over the course of my career, we've heard less and less about him. When I was a student in the 1970s, we still had many lectures about Freud, people still talked about him a lot, often to disagree or to say he was wrong about things. But he was still an important presence. Even when I started working as a professor, there were those who really used Freud as the foundation of their clinical practice, who thought he was the foremost thinker. He certainly did advance the field. I think what happens is that the things someone like Freud is right about get taken over gradually by other people, and so they cease giving him credit for them. They start to focus mainly on the things he was wrong about, and so they were gradually discredited and forgotten, so nobody in psychology talks about Freud anymore. Maybe I wouldn't say "nobody," but I read a lot of psychology, and I go to conferences to talk about things. It's very rare to hear anyone mention Sigmund Freud anymore.

So that approach has faded away, but it gave some inspiration to me in terms of meaning of life. Freud was a big thinker, saying things like "love and work would be the two main things." I periodically go back to that and think about meaning of life. Freud was very skeptical of religion. To some people, obviously, the meaning of life is not religion, but there aren't a lot of other things on top of loving and working. I'd say Freud shaped my thinking. I don't go back and reread or cite him, but he still has a lasting influence on my thinking. The Bible—not so much.

For me personally one of the most shocking results of your research is how it demonstrates that psychology and social psychology in particular has become a science of self-reports and finger movements (Baumeister et al. 2007)**. I expected you work to shock the world of researchers, especially considering how, in the middle of the decade, it was proclaimed as the decade of behavior. Was this in fact the case? Did the scientific community perceive it as a cold shower, or did they remain silent?**

The paper's been cited over a thousand times, so a lot of people noticed it and read it. I've heard many people agree with it and many others say that it was wrong. Recently, several other articles have come out saying it's intensified. I'm not sure how much you're up on what's going on in psychology lately, but there are many concerns about replication and statistical power and so on which have led people to say "okay, everything has to be giant samples now and the only way to get giant samples is to do these big online surveys." As a result, all of what social psychology does anymore is people sitting at computers making ratings, and personality psychology's not much better. Although they have started to map out and go into big data and look at who becomes more successful, who lives longer, and who earns more money. So, personality at least is moving beyond that, but social psychology is the science of finger movements, and it's getting worse.

The Mechanical Turk online sample and the outsourcing in recent articles has become 90% of what's being done in our top journals. It's a long way from where social psychology started with the Milgram studies, Schachter and Singer's work on emotion, and the Darlene-Latané bystander intervention studies. A lot of effort was put into staging an experiment. You ran people one-on-one, the participant would come to the laboratory for an hour, and you set up an elaborate experience. You manipulated this way and that and observed actual behavior.

I want to emphasize that there's nothing wrong with the ratings and the finger movements. We learn from them too. But for that to be the only thing we do, I think we are losing a lot. A scientific field is strong when it has multiple diverse methods. There's no perfect method in the social sciences. If you want to learn that something is true, the best we can do is to get converging evidence from multiple methods. So when we start to all fall back on the same method all the time that weakens the field. The way I look at social psychology, in the 1970s when I went into the field it was rather narrow. There were only a few methods and a lot of work was done on just a few topics. So it was really not that intellectually rich a field.

I remember how disappointed I was in graduate school, especially coming from philosophy, where the ideas are much more developed and conceptually sophisticated, and I thought, "I'll just have to make do." But then in the 1980s there was this great methodological expansion. And we need to study more different things. Since we can't do an equally good job of everything, we need to adjust to study, for example, social aspects of memory, since we have excellent methods there that have to be just perfectly done, everything has to be right. But if we study marriage, we can't randomly assign people to be married or track them over the years. We have to lower the standards, but we can't *not* study marriage, that's no solution. Today the thinking seems to be "no, we should not study anything that we cannot do with these big online samples." I think that's going to be a weakness in the field. I'm less optimistic about the field than I have been at other points throughout my career.

About a year ago, a Polish scientist, Dariusz Doliński (2018)**, replicated your research. And the *Social Psychological Bulletin* in which its results appeared invited researchers to engage in discussion and**

devoted an entire issue to this subject. Have you read his work and the discussion it evoked?

Yes, I read that one. In fact, they may have even sent it to us to review, or they asked if we wanted to write a comment. I consulted with my co-authors and we were glad to support what he was doing, but we didn't really have anything new to say about it ourselves. But yes, his is one of the papers saying that the situation has gotten worse.

Can the abandoning by psychology of behavioral studies and the turn toward introspection be considered one of the causes of the crisis in psychology discussed so openly in recent years?

Both those things are happening, but I don't know that introspection is the cause of it. I think the crisis probably had different origins.

What, in your opinion, are the fundamental causes of this crisis?

I don't think there is cause for people to question what we're doing in general. There were the cases of fraud. Diederik Stapel was a particularly disturbing one because it wasn't just one person doing one fake study here and there. He was an established researcher who did a lot of fraud, he won awards and was very productive, people looked up to him. Some people are faking data. Another source of the crisis, without fraud, was Daryl Bem (2011), who published an article on precognition, a sort of extrasensory perception. Some people just say there cannot be any such thing. So the fact that he could get these results shows that our research methods are flawed. I'm not sure I agree with that argument, but I see their point.

And then some people started doing replications. There's one by Nosek, a hundred studies or something like that. His group ran a bunch of them. But then many of them didn't work, which again increased the sense of crisis. People have got the idea that everything should replicate all the time as if there are laws of psychology. Replication is a concern. Still, everyone presumably learned in methods classes that a null finding is inherently ambiguous.

For this younger generation of researchers to think "if something doesn't work out for me it can't be my fault, it must be somebody else's fault," that's probably the product of the self-esteem movement. My feelings are ambiguous.

I'm not a historian of science, and I don't know why some things just come together, but I know there are things that do replicate and things that don't. We replicate other people's work, and often we get a finding, but sometimes we don't. I think the modern thing is to say "it must be them, they must be faking their data like Diederick Stapel," but I don't think that way. I'm more of a literature reviewer than most scientists. I write a lot of large literature reviews, I read a lot of other people's work, and I cite it. So I have a more positive attitude toward social scientists in general than other people. I know other scientists, even some of my friends, who are hardly interested in anything other than what they and their friends are doing, and they think what everyone else is doing is stupid and wrong. Fine, then they're not going to write literature reviews. But I am, because I want to see the big picture. I have to believe that most social scientists are decent people trying to do a good job and find the truth.

To me, the biggest danger is political bias, not erratic replication. A lot of social scientists have strong political views, and they view their research as a way of advancing social justice. In my view, it's hard to do that and pursue the truth at the same time. I'm not into political views and social justice and all that—I just want to end up knowing what's right, and I don't mind revising my opinions. If you have a political investment, this is like having religious beliefs—but most social scientists aren't religious. If you're invested in some political view, then you don't want to find out any contrary facts, and you want to set up the experiment to get things your way. I think most social scientists are doing reasonable work as best they can and trying to find the truth, with the exception of the people who have strong political biases. I think that's the biggest threat to the integrity of the field, more so than the replication issue.

When discussing issues related to the crisis in psychology, we must keep in mind the doubts that have arisen from the reputation of the research that underpins your ego depletion theory (Baumeister et al. 1998)**. Do we have more data today to dispel these doubts more unambiguously?**

I recently wrote something on that which thousands of people have read (Baumeister 2019). Ego depletion been found over and over again on multiple continents and in multiple different samples and contexts.

Well, some people haven't got it yet, perhaps it's an issue of timing. But by the time two or three different independent labs get the effect, there has to be something real there. I think the people who are trying to discredit the whole field think ego depletion is a good thing to attack because it's one of the best replicated findings. They could say "well, there is something that's been found several hundred times, but it's still wrong." If they said that it would go a long way toward discrediting everything in the field. But to me that's ridiculous. Something that's been found several hundred times has got to be real. There's no way that there could be a conspiracy to produce that much fake data, nor could it be capitalizing on chance or any other source. The theory could be wrong, but there is definitely a real phenomenon there.

The notion it was capitalizing on chance was more or less devastated by that paper by Friese (Friese et al. 2019), who pointed out that if it were chance, there would be just as many significant findings in the opposite direction. So you have around six hundred significant findings showing that depletion makes you do worse things. There should be an equal number showing that depletion makes you do better things, to demonstrate capitalizing on chance. Journals would be eager to publish those because they will have greater novelty, but there are only two—one of which is mine, and one by other researchers. They're both sort of interesting exceptions. So that idea that it's some kind of chance or distortion of public publication bias is preposterous. I honestly can't understand how any sincere honest scientist could question the reality of ego depletion.

Looking at what you do from a broader perspective: Is there any particular type of criticism aimed at you that you feel is particularly serious and justified?

The best alternative explanation for ego depletion was not a complete alternate explanation. It's an alternative to the resource depletion idea. It's mentioned in an obscure article by some researchers from upstate New York (Ampel et al. 2018). I reviewed it for several other journals and argued that they should publish it, but they didn't. The journal usually get a brain person saying "oh, we already know all this," so they rejected it. But after a time it finally got published somewhere. It's a really interesting idea, that it's not that the brain has used up too much glucose or other fuel, but rather that maintaining high glucose in a certain area is bad for the neurons. For example, people with diabetes eventually lose sensation in their feet because of high levels of glucose. If you keep high levels of glucose over a long period of time that damages the nerves, and for that to happen in the brain could be very bad. I know it's the case that alcohol kills brain cells too, and the brain is pretty adaptable, but if you somehow burned out all the nerve cells in your self-controlling area that would be bad. So maybe the reason people show depletion is that they've just exerted self-control. They want to let the cells cool off, so to speak. It's not running out of resources, but rather the brain wants to let the level of glucose diminish to avoid harming the cells. I think that's the best alternative theory that I know.

In terms of the work on belongingness and rejection, I mentioned the idea that your self-esteem is about hierarchy, not just acceptance and belongingness. Sedikides and Gebauer (2014) have put forward this theory. I think that is a valid point and I've changed my opinion to incorporate that view.

Your question's a bit difficult for me to answer because I never start off thinking I know what's right. So I read other people's criticisms and I say "oh, yeah, they're right," and I include that into my own thinking. This happens a lot. In terms of where we are now, I learned there was a paper criticizing the glucose idea about 10 years ago, saying it's really not about running out of glucose. It's more about allocating it because the body has plenty of glucose stored. And I thought "that's completely right." So I've changed my thinking to adopt their theory.

They thought you could omit the limited resource notion. But I think you usually have selective allocation for a resource that is precious, and when a resource really is unlimited then you don't need to allocate it carefully or selectively. There has to be some cost or some problem for allocation to be selective. Like I said, my theories are constantly being revised, and when someone comes along with a serious point or criticism that I agree with, I adopt it. I have this body of work on the theory of prospective thinking, and there's been some serious criticism of that.

I talk a lot about sex and gender. People object to that, but they often seem to be politically motivated. The issue of sex and gender is so politicized that I'm thinking I don't really want to work on that area. It's too hard to be open-minded, to just look at the facts from all sides. There is a sort of societal commitment to believe a bunch of things that are ridiculous when looked at as scientific hypotheses, but anybody who dissents gets hunted and denounced, and is subjected to character assassination. To me as a literature reviewer that's a sign, when people who raise contrary views are attacked and silenced. As gender goes, the field has had journal articles published and then retracted based on political pressure. These are clear instances of censorship, which is a red flag that there's no intellectual freedom there, and I think that's unfortunate.

I know that in America issues of race are very carefully guarded too, but for my project I'm trying to understand what human life is all about. I can dispense with race to an extent because most interactions are with one's own race, and the interracial things are somewhat marginal, but I can't do without gender. If I want to really understand what human life is all about, male, female, or undecided, that's a big part of everyone's life. So I really had to delve into that.

What research problems you are trying to solve to these days, the issues you are presently engaging?

These days I'm writing a book on the self. I'm trying to pull together what I've learned in many different areas of self over the years, self-esteem and self-control, and other things like self-presentation. I hope to have a first draft done by the end of the year. In terms of new data and theorizing, I have been thinking about the future, this has been an interest of mine in recent years. I recently had a grant application turned down, which was for how people think about the future. That issue goes to some of the most basic aspects of the human mind and the ways in which our minds are superior to those of other animals. It goes along with understanding the self, understanding consciousness and free will and all those things. Linking across time, narrative understanding, mapping out the multiple possibilities in the future rather than just trying to predict the one thing that's sure to happen. That's been a lively and interesting field of study.

I have various small projects with scholars in different places. These are moving along at different speeds. We just had one paper accepted on soccer, based on the World Cup (Nezlek et al. 2019). That was sort of an intriguing little project during the 2014 World Cup. They surveyed people in different countries about how much corruption there was in FIFA, and it turned out that countries which had more corruption themselves saw less corruption in FIFA. The ones who were the most outspoken critics of FIFA being corrupt were the countries that were lower in terms of corruption, they're shocked by it. For people who live in countries where you always have to pay bribes and everyone's corrupt, the idea that the football organization was corrupt didn't really seem to bother them that much. But countries where you don't pay bribes and so on were shocked at this. I'm doing a variety of smaller things like that.

My thought moving forward is to write big literature reviews and books, as I'm getting older now and I don't have that much time left. I want to tackle the big questions, so I'm also cutting back on working with Ph.D. students. You need to take care of them and have them do narrowly focused studies, then try to publish them so they can get jobs. You end up working with small questions and writing things up. Some papers are just "good" but there's pressure to publish, even if things didn't work out so well you want to salvage something from it. I think I don't want to do as much of that, and in my remaining time, I want to do the big picture thinking, try to pull together grand issues. I've got several ideas for books to do on cultural change, and I wouldn't mind actually doing another meaning of life book. The one I did in 1991 is still selling pretty well and it's my most cited book in the scientific literature, but there's a lot more evidence to work with now. Presumably, I'm more mature as an intellectual, so it would be interesting to revisit those ideas and questions and write an all new book on that topic.

Sounds interesting. I would like to discuss with you some more general issues. In your opinion, what crucial questions have psychology managed to answer so far?

That's a hard one. A conversation I had with Richard Wiseman comes to mind. He was going on a panel with people from different disciplines, and they said they were going to ask everybody "what has your field really learned in the last ten years? What's the biggest advance?" They wanted us to say what we thing psychology's biggest advance in the last 10 years is, but there's so many different things in psychology going on that it's hard to pick a big one. We had trouble coming up with anything, and when I saw him again the next day I said "did you think of anything?" He said "yeah, I think what I'm going to say is what we've learned last 10 years is how little difference the brain makes." I thought that was really quite a profound idea as we've spent enormous amounts of money on studying the brain and we've learned a lot about which part of the brain is relevant to this or that, and yet it changes our psychological theories very little.

This is something I've noticed for a long time. I remember exchanges where we had trained experts come to the summer institute where I worked, and they would talk about what we learned about the brain, and other people would say "how does that change how we think about marriage or people's attitudes?" And he said "well, it doesn't, really." I don't want to put words in Wiseman's mouth, but my take is that the brain is basically responding to its environment. It's not an independent cause, it's not the root cause.

I think many psychologists had the idea that once we understand the brain, we won't need all the rest of psychology. And that is completely wrong. Even if we understand perfectly how the brain works, it's just responding to what's happening in the environment and we'll still need to understand marriage and aggression and helping and all those other things. The brain is just a mediator that really can't even afford to change things very much. It more or less has to see the world precisely as it is, because distortions will be maladaptive and you'll react in the wrong way. You've got to see what's really there and deal with it.

Some of the big changes in my lifetime has involved the shift from the socialization and Freudian theories of the 1960s to a much more biological approach. Yet that was oversold too. For a while, people were saying "everything is genetic," but in fact genes are just one part of it. The gene-environment interactions seem to be by far where all the action is. We are born with strong predispositions to act in certain ways, but the environment can turn some of them on and off and change them, and they don't usually revert.

I could talk about gender differences that men and women might start out with certain different inclinations. Most societies in history would have exaggerated those and steered men and women into completely different lives. Our modern society is the other way around, trying to eliminate the differences and treat boys and girls the same, teach them to be the same and so on. Society can do that, make natural tendencies bigger or smaller. It doesn't usually reverse them. There are very few societies where gangs of middle-aged women are roaming the city, getting into fights, beating people up and burning things down and so on. These reversals don't happen. But I think understanding how biological nature and social situations work together has been a big advance. I think a lot of the advances of psychology are at a much more specific level, and I think we've learned and done a lot of things. But as for grand problems, I'm not so sure.

Which serious questions in our field still remain unanswered?

Consciousness is still very difficult to understand. What the philosophers called the hard problem of consciousness is how physical items like brains can produce subjective conscious experience. We have almost no idea about that. I think we've learned a lot about evil and violence and helping. I'd say where we are is not that questions are totally unanswered, but that we have tentative answers for many things. Some of these answers are better than others, and they're likely to last longer than others.

We've certainly learned a lot about male and female. We're distorting it because of political correctness biases and those things, but nevertheless we have certainly learned a lot. We'll see just how durable those things are in the future. It's not so much that questions are unanswered, but some of the answers are wrong.

You mentioned that we have many tentative answers, and in general, psychology as science is not a very consistent system of knowledge. Do you think we will finally achieve a wide-ranging theory that incorporates the achievements of psychology into a uniform system of knowledge?

Not in our lifetime, that's for sure. I don't see any of the social sciences moving toward an integrative theory. I think there's good progress. There's lots of little pieces and I would like to think that someday it will happen, but let's go back to all the studies with rats, the place where psychology was just before I came into the field. We learned a lot from reward and punishment, partial reinforcement schedules, and all those things. That's pretty basic and valid—but the idea that it would explain all of human behavior was ridiculous. Still, it was definitely some true and valid knowledge.

Now, in the psychology of people there is the issue that people may change. How much people can change is an open question but, for example, people today are much more tolerant and less aggressive than in the past. Change has to happen but change then makes theories obsolete. There might be some true theories that will then cease to be true in the future. But there should be some commonalities.

What is your opinion about artificial intelligence? Do you think it will help us to understand human beings somehow, or it's irrelevant, or something else?

I have not really followed it very closely. My approach as a literature reviewer is that usually there's some valid points and they are usually overstated. Artificial intelligence is a clear example of that. Certainly, we've learned some things about human intelligence from it, but it's not going to provide all the answers by any means just by virtue of being artificial. The way a computer thinks is not necessarily how people think. When a computer plays chess, it really thinks through every possible move and all the contingencies. It can compute those because it has great power, but the human mind can't do that. The human mind plays chess by some other procedure. Perhaps computers can model those as well, but I would say we'll see limited benefit from artificial intelligence in terms of psychology. It's also a tool. Artificial intelligence can do things that real human intelligence would not be able to do. Computers can combine vast amounts of information and sort through it, finding patterns in a way that would be almost impossible for a single human mind to do, so they can make life better in lots of ways. But they're a tool for us. As a psychological tool, it can teach us some things about the human mind but not everything.

Many young people at the threshold of their careers are looking for advice on what direction to take. What would you tell them?

It's unfortunate you're asking me this now because for many years I mentored a great many young people and gave them lots of advice. Around 40 people got their research careers started by working with me. Today, I'm not as confident about the future of the field and I don't know what the right thing is. I don't even know what field I would go into if I were starting out with the benefit of my own wisdom.

One good bit of advice is while you're in graduate school to take all the statistics you can, because in the long run that limits the questions you can ask and answer. I tell my students I had my last statistics course in 1976. And nobody takes statistics courses after you're a professor. So I get things to review where they've got statistical analyses with some methods that weren't even invented when I took my last class. You want to get as much on top of things as you can because you want that statistical knowledge to sustain you for the decades to come.

Then, I advise them to practice writing skills, to get to where you write every day. Writing hard for everyone at first, but it gets easier as you continue to do it, and the automatization process works so that your writing gets better.

Let's say you've got to learn multiple methods, lab procedures and statistical approaches, that work for you while you are still in graduate school, so that when you get a job as a professor, you can just capitalize on those things. The first few years while you earn tenure you need to produce, so that's not a time to be exploring new methods. You explore new methods when you're earning a Ph.D., and then after you get tenure, you can explore new methods and try other things—that's what tenure's for. But to earn tenure you have to play the game by the rules, the way other people do it, and you have to have something already that works for you that you can reliably produce interesting results on. I would tell them all those things: Learn how research is done and do that, prove yourself first, then you are entitled to go off on your own and tackle grander things.

Quite practical. Would you encourage young people to choose to swim upstream against the dominant tendencies of the time?

Perhaps not entirely. I think you want to keep that in mind and maybe do a little of that. But remember, you have to earn tenure based on the methods and standards that prevail at the time. So when you start out you have to do something that will get you published and earn the approval of the decision makers. Once you've done that, then by all means also do some swimming upstream and question established ways of doing things. But by doing everything the opposite of what everyone else is doing, you'll probably just get frustrated and you won't get funded. It'll be difficult to get tenure and you'll end up dropping out. Establish your career first. Once you have tenure, you can do whatever you want, that's its purpose, so people can challenge existing ideas. But to get there you have to prove yourself by what the field thinks is important and the right way to do things, and I think that's not a bad model at all.

Surprise question if you don't mind.

Sure.

I have a list of 30 questions that my readers would like to ask the most influential psychologists in the world. Please draw one.

Give me number one.

Have you ever cheated as a scientist?

No. The question surprises me, because it seems to assume that if you asked someone like Stapel, who cheated extensively, he would say yes. I imagine that fraudsters (and my view is there are probably very few of these among today's scientists) would deny cheating, thus giving the same response as honest people. I think Stapel's case is revealing, because he started cheating early and then did it frequently. Once a scientist is established, there is far more to lose than gain by cheating, so even if someone lacked moral scruples, he or she would still not take the risk of engaging in any sort of fraud.

To be sure, norms change. I recently spoke with one of my mentors, now in his 70s, and he remarked that what is now despised and condemned as p-hacking was once considered best practice. That is, if your study did not yield significant results, it was reasonable to try various other analyses, such as using a covariate to get rid of some of the error variance. This might nudge a finding across the .05 significance level. Today, that might be considered cheating. Back in the 1970s, when I was being trained, we were all taught to do that. I think most of us have changed to go along with the new norms (I certainly have), but if you apply today's norms to judge older work from the 1960s and 1970s, you might say that almost everyone cheated back then. I don't agree with that view, judging past work by today's standards, but others may disagree.

Personally, I had a pretty strict upbringing, so truth, honesty and loyalty are strong feelings for me. For others, I can understand cheating when you're starting out, because you just want to get something published. But by the time you've gotten tenure, you've published a bunch of real work and anything dishonest would cast all your work into doubt. I was talking to another professor here yesterday. We work with many people, and we discussed what if one of them cheated, what if one of them faked data. We can't check all their work and you have to trust the people you work with. The conversation was stimulated by a case in another laboratory, in which someone who trusted someone else who then did something wrong and it was bad for everyone. You live in fear of that happening. I've got around 660 publications, and to think if someone came into my lab and faked data and it were found out and I didn't notice it, I believe it that would cast doubt on all those 660 publications. It's an ongoing fear of scientists that someone in a lab might do this.

As science becomes more and more collaborative, we rely more on other people to tell the truth, and then one person can do an enormous amount of damage. Again, going back to Stapel—he not only ruined his own life, but many other people's lives because they had published with him. Students and proteges found even their honest work came under suspicion and was tainted. Imagine, some of them are just about to come up for tenure and suddenly half the papers that they put out are under suspicion. They did things honestly, but just by virtue of being associated with Stapel it was all under question, and they couldn't use it as a credential. There's a whole lot more that could be said about this.

Selected Readings

Baumeister, R. F. (1986). *Identity: Cultural change and the struggle for self*. New York: Oxford University Press.

Baumeister, R. F. (1989). *Masochism and the self*. Hillsdale, NJ: Lawrence Erlbaum Associates.

Baumeister, R. F. (1991a). *Meanings of life*. New York: Guilford Press.

Baumeister, R. F. (1991b). *Escaping the self: Alcoholism, spirituality, masochism, and other flights from the burden of selfhood*. New York: Basic Books.

Baumeister, R. F. (Ed.). (1993). *Self-esteem: The puzzle of low self-regard*. New York: Plenum.

Baumeister, R. F. (1997). *Evil: Inside human violence and cruelty*. New York: W.H. Freeman.

Baumeister, R. F. (2005). *The cultural animal: Human nature, meaning, and social life*. New York: Oxford University Press.

Baumeister, R.F. (2010). *Is there anything good about men?* New York: Oxford University Press.

Baumeister, R. F. (2018). *Self-regulation and self-control: Selected works of Roy F. Baumeister*. New York: Routledge Psychology Press.

Baumeister, R. F., Heatherton, T., & Tice, D. (1994). *Losing control: How and why people fail at self-regulation*. New York: Academic Press.

Baumeister, R.F., & Tierney, J. (2011). *Willpower: Rediscovering the greatest human strength*. New York: Penguin Press.

Baumeister, R. F., & Wotman, S. R. (1992). *Breaking hearts: The two sides of unrequited love*. New York: Guilford Press.

Berglas, S. C., & Baumeister, R. F. (1993). *Your own worst enemy: Understanding the paradox of self-defeating behavior*. New York: Basic Books.

Seligman, M., Railton, P., Baumeister, R., & Sripada, C. (2016). *Homo prospectus*. New York: Oxford University Press.

Tierney, J., & Baumeister, R.F. (2019). *The power of bad: How the negativity effect rules us—And how we can rule it*. New York: Penguin.

References

Ampel, B. C., Muraven, M., & McNay, E. C. (2018). Mental work requires physical energy: Self-control is neither exception nor exceptional. *Frontiers in Psychology, 9,* 1005.

Baumeister, R. (2019, September 4). *Self-control, ego depletion, and social psychology's replication crisis*. Retrieved from https://doi.org/10.31234/osf.io/uf3cn.

Baumeister, R. F. (1990). Suicide as escape from self. *Psychological Review, 97*(1), 90–113.

Baumeister, R. F. (1998). The self. In D. T. Gilbert, S. T. Fiske, & G. Lindzey (Eds.), *The handbook of social psychology* (4th ed., Vol. 1, pp. 680–740). New York, NY: McGraw-Hill.

Baumeister, R. F., Bratslavsky, E., Muraven, M., & Tice, D. M. (1998). Ego depletion: Is the active self a limited resource? *Journal of Personality and Social Psychology, 74*(5), 1252–1265.

Baumeister, R. F., Campbell, J. D., Krueger, J. I., & Vohs, K. D. (2003). Does high self-esteem cause better performance, interpersonal success, happiness, or healthier lifestyles? *Psychological Science in the Public Interest, 4*(1), 1–44.

Baumeister, R. F., & Leary, M. R. (1995). The need to belong: Desire for interpersonal attachments as a fundamental human motivation. *Psychological Bulletin, 117*(3), 497–529.

Baumeister, R. F., Smart, L., & Boden, J. M. (1996). Relation of threatened egotism to violence and aggression: The dark side of high self-esteem. *Psychological Review, 103*(1), 5–33.

Baumeister, R.F., & Tierney, J. (2011). *Willpower: Rediscovering the greatest human strength*. New York: Penguin Press.

Baumeister, R. F., & Vohs, K. D. (Eds.). (2004). *Handbook of self-regulation: Research, theory, and applications*. New York, NY: The Guilford Press.

Baumeister, R. F., Vohs, K. D., & Funder, D. C. (2007). Psychology as the science of self-reports and finger movements: Whatever happened to actual behavior? *Perspectives on Psychological Science, 2*(4), 396–403.

Bem, D. J. (2011). Feeling the future: Experimental evidence for anomalous retroactive influences on cognition and affect. *Journal of Personality and Social Psychology, 100*(3), 407–425.

Branden, N. (1969). *The psychology of self-esteem*. New York: Bantam.

Doliński, D. (2018). Is psychology still a science of behaviour? *Social Psychological Bulletin, 13*(2), retrieved from https://doi.org/10.5964/spb.v13i2.25025

Friese, M., Loschelder, D. D., Gieseler, K., Frankenbach, J., & Inzlicht, M. (2019). Is ego depletion real? An analysis of arguments. *Personality and Social Psychology Review, 23*(2), 107–131.

Hood, B. (2012). *The self illusion: How the social brain creates identity*. Oxford: University Press.

Joiner, T. E. (2005). *Why people die by suicide*. Cambridge, MA: Harvard University Press.

Jongman-Sereno, K. P., & Leary, M. R. (2018). The enigma of being yourself: A critical examination of the concept of authenticity. *Review of General Psychology, 23*(1), 133–142.

Nezlek, J. B., Newman, D. B., Schütz, A., Baumeister, R. F., Schug, J., Joshanloo, M., et al. (2019). An international survey of perceptions of the 2014 FIFA World Cup: National levels of corruption as context for perceptions of institutional corruption. *PLoS ONE, 14*, e0222492.

Sedikides, C., & Gebauer, J. E. (2014). Religion and the self. In V. Saroglou (Ed.), *Religion, personality, and social behavior* (pp. 46–70). New York, NY: Psychology Press.

Storr, W. (2014). The man who destroyed America's ego. *Matter,* retrieved from https://medium.com/matter/the-man-who-destroyed-americas-ego-94d214257b5.

12

Erica Burman: Developmental Psychology, Feminist Research and Methodology

It is a fundamental misunderstanding to see being critical as negative. Critique is vital for renewing and changing ideas and practices.

Erica Burman

T. Witkowski, *Shaping Psychology*,
https://doi.org/10.1007/978-3-030-50003-0_12

Erica Burman was born in Liverpool, as the granddaughter of Ashkenazi Jewish immigrants from Eastern Europe. She has lived in Manchester since her doctoral studies, and those community and social links have been important resources supporting the activist research she has been able to undertake. When telling her story, Erica Burman says she never planned to become a critical psychologist, nor to take up the field of feminist psychology. As an undergraduate she got involved in cognitive studies, an area dominated at the time by computational models of mind and the beginning of AI. She started to grasp the extent to which science is at times dominated by politics. At the time she was also studying philosophy of the mind, which—via its interrogation of terms, language and logic—led her to take a critical perspective on psychology. Another source of inspiration for skeptical analysis of what she found in psychology was the range of professional practitioners she taught when she started to teach developmental psychology. As she puts it, it was from them that she learned to critically think and look at the practice of psychology. To this day she feels that applied psychology should be continually analyzed in terms of how much it comes to people's aid, and how much it constitutes an extension of oppressive social practices. In defining herself as a critical and feminist psychologist, she emphasizes at the same time that these are not separate sub-disciplines of our field, and does not think it necessary to create them (since arguably critical psychologists try to dismantle rather than build psychology). It is simply an attitude or posture that makes it possible to look at the theory and practice of psychology differently.

One of her more important and influential works is the 1994 book *Deconstructing Developmental Psychology*, which has now seen three editions published (most recently in 2017, with a second edition of her *Developments: Child, Image, Nation* now in press). *Deconstructing Developmental Psychology* is a critical evaluation of mainstream theories of child development and drew upon feminist theory to show how this aspect of psychology serves to regulate family behavior, marginalize working class and minority ethnic women and pathologize their experience as mothers. The book covers the spectrum of dominant approaches in psychology and finds each of them wanting. Erica Burman examines in the book specific cultural assumptions that give rise to different forms of psychology and provides new ways of thinking about the situation of children in contemporary society.

Many later studies by Burman have been devoted to representations of children and to the relations between different kinds of development (hence the second book, *Developments: Child, Image, Nation*). Continuing critical examination of the role of developmental psychology, she turned to study the way images of children are used in connection with the "developing" world. She has also focused her research on the question of how such images of women and children hold in place models of the "progress" of the development of the nation state and directed several national and transnational projects on service support for minority ethnic women facing state and interpersonal violence (informed by these perspectives). Her current work explores postcolonial and decolonization perspectives as they inform ideas about children and development in the sense of both individual and economic development—that is, as practiced in national and international policies and in pedagogies.

Burman is not only a declared feminist psychologist. She also has questioned cultural assumptions in second wave feminism and has often drawn on anti-racist debates. She has directed attention to the way the position of women is closely connected with the position of cultural minorities. Burman is best known as a developmental psychologist and theorist of women's studies, but many of her publications have also been concerned with radical developments in methodology. Her study of different ways of carrying out research has been a powerful resource for feminist psychologists, but beyond that the impact has been felt in discourse analysis, critical psychology and critical mental health.

Erica Burman works also as a group analyst, and much of her work has been done in cooperation with a new generation of researchers working in psychology and in similar disciplines, as well as with practitioners. Among the many books co-written and edited by Erica Burman are *Challenging Women: Psychology's Exclusions, Feminist Possibilities* and *Psychology Discourse Practice: From Regulation to Resistance*, practical expressions of this critical feminist work.

In terms of collaborative research projects, she has devoted much energy to such issues as suicide and self-harm, domestic violence, challenging the racism of immigration controls and the ways poor communities are further stigmatized by cuts in welfare support. Burman was a co-founder (with Ian Parker) of the Discourse Unit, and the website of this research unit provides open access to a number of publications.

Probably the first significant critical trend in our field which employed deconstruction as a method on a large scale, although probably not in a very systematic manner, was anti-psychiatry. How do you, as an adherent of critical psychology and a supporter of deconstruction, evaluate the achievements of this movement in retrospect?

To be honest, I wouldn't have identified anti-psychiatry directly with deconstruction, but it's an interesting connection. As a critical feminist psychologist, I have some mixed feeling about anti-psychiatry, not least that—since it was always led by psychiatrists—it paradoxically reinforced the role of psychiatric (male) authority. It therefore reconstructed, as much as deconstructed, psychiatry—albeit also helping usher in some other discourses and practices. The democratic psychiatry movement inspired by political developments in Trieste, Italy, while not maintained entirely, certainly showed how coercive institutions produce the phenomena they need to regulate. Hence if the threat of coercive treatment or incarceration is reduced, then the need for those measures is dramatically reduced. This goes for medication practices too.

In your opinion, did the activity of anti-psychiatrists also have any negative effects?

Lots. In dominating the domain of critiques of psychiatric models and practices, it was rather weak and limited, and many key proponents still subscribed to very normative and individualist models of mental health and the organization of social relations (as in labor, etc., individual responsibility).

Just a few weeks before our conversation, the premiere of Susannah Cahalan's book *The Great Pretender* (2019) took place, which casts doubt on the veracity of the experiment by David Rosenhan involving sending pseudo-patients to psychiatric hospitals in the USA. This high-profile experiment commonly associated with

the anti-psychiatry movement has ushered in revolutionary changes in the treatment of mental illness. What do you think about the discoveries made by Cahalan?

Well the psychoanalyst Jacques Lacan says "truth has the structure of fiction"... So something made up can still be informative... But of course this speaks to a different kind of truth from the one that positivist empiricists like. Of course it was useful to refer to Rosenhan's accounts but such anecdotes as the ones he wrote about have wider resonance which is why it was so influential. In that sense, it doesn't really matter so much. And it could be said that experiments, too, have the pedagogical structure of parables in the style of their narration...

Can we draw any constructive conclusions from the experience of the anti-psychiatry movement as to future activity involving the application of deconstruction?

When we ran day conferences at our university in partnership with the Hearing Voices Movement, inspired by the work of the Dutch social psychiatrist Marius Romme (Romme and Escher 1989), we saw this as a form of practical deconstruction. On the panel, speaking about the experience of hearing voices was a psychiatrist, a clinical psychologist, but also a shaman and a spiritualist. The accounts of experts by training (all trained in very different expertise) and experts by experience (including psychiatric system survivors) were situated alongside each other, with no single one privileged. Instead each was one relativized, and the dominant accounts (such as the psychiatric one) were correspondingly destabilized from its hegemonic position and shown to be merely one position among various. This shows how deconstruction is a practical activity undertaken collectively by people, including those oppressed by practices of power; and how this practical activity can produce change in models of mental health practice—as indeed the Hearing Voices Movement has done, worldwide.

At this point in our conversation, the question about your attitude to psychiatric categories of mental illness and diagnostic textbooks such as DSM or ICD arises.

I see these as historical documents that testify to the cultural and political preoccupations of their times (think of how "homosexuality" was only removed as a "pathology" in 1973, and by postal vote). Also reflecting prevailing orientations, the early editions of the DSM were psychoanalytically-informed, the later ones behavioral, and now there are alignments neuroscientific accounts. In general, I am suspicious of the tie-up between psychiatry and the pharmaceutical industry. Psychiatry is a branch of medicine and it is regrettable that it dominates psychological thinking. Personally, I do not find the psychiatric labels helpful or convincing (even if some people so labeled find some comfort in those labels, as we discuss elsewhere in this interview), and the medical model can disempower people and be used to deny their human rights, while the so-called treatments can disable people from being able to deal with their difficulties. I would much prefer to talk of distress or suffering than mental illness, though as Frantz Fanon argued even where distress is produced by social conditions (of oppression or marginalization) this will still be experienced in specific ways and need to be engaged with in the light of individual, idiosyncratic associations, perceptions and circumstances. In relation to the status of the DSM, my views are certainly not atypical. The Division of Clinical Psychologists (the group of professional psychologists who work most closely on the terrain of psychiatry) of the British Psychological Society have refused to adopt the DSM V, and in 2018 put forward their own approach of formulation instead. While not official policy of the whole Division, they also published an important document that offers a coherent alternative model of and approach to working with distress (or mental illness/pathology), which they call the Power, Threat, Meaning Framework. This sees distress as a response to threat experienced in unequal or oppressive conditions, and works with the meanings of these experiences via narrative approaches. For the authors of that document, this is a progressive psychological intervention. For me, it indicates a way out of psychology altogether.

I started our conversation asking about these historical events because the critical approach and method of deconstruction are still poorly known, sometimes distorted, and when referring to quite well-known events I am convinced that readers will find it easier to understand what we are talking about. But let's move on to contemporary issues. When I observe public debates regarding issues involving children, I have the irresistible impression that they are almost always treated as someone's property. Some parents refuse to

vaccinate their children, claiming that nobody can impose decisions as to what they do with them. On the other hand, the state sometimes takes on the role of a policeman who takes "property" away from parents in order to manage their well-being as they see fit. The same applies to education. Decisions are always made for them by someone else. Do you think that children have any chance of any kind of autonomy at all? What does critical psychology offer in this respect?

I cannot speak for critical psychology, as this means different things in different times and places (necessarily so, as it is constructed in relation to other kinds of psychology and social practices operating in a range of socio-political conditions). In relation to this specific question, as someone working on the border of developmental psychology and childhood studies, these are vital and urgent considerations but not ones that are easily answered in simple binary terms (of yes or no). But here goes. Of course children exercise forms of autonomy and need to be recognized as doing so. Quite how they do so, and how they should be enabled to do so, requires some complex thinking about our models of how agency and autonomy—as well as functioning as significant ideological tropes—are actually practiced in relational and interdependent relationships. In this sense, I see this as a general question for us all, not just for children—even though children may function as a key expression of such relational and interdependent dynamics and relations (as proxies, property, or as sentimentalized, nostalgic representations of our lost selves, etc.). This is why I have formulated the approach I call "child as method" to see theories and practices around children and childhood as a way of diagnosing wider sets of issues and relationships at play (Burman 2019a).

Could you say something more about the "child as method" approach? This is a quite new approach and I am afraid that perhaps not all of our readers are familiar with this.

I'm sure your readers won't be, as it is a new framework that I have generated recently. I draw upon recent postcolonial and anticapitalist analyses, such as Kuan-Hsing Chen's (2010) *Asia as Method*, and Sandro Mezzadra and Brett Neilson's (2013) *Border as Method* to offer an analytic framework for considering the relationships between models of childhood and other sociocultural and historical axes produced under particular geopolitical conditions. This is an intervention both from and in childhood studies, bringing insights from social theory to interpret the position of children, and the meanings accorded childhood, as not only significant issues worthy of attention in their own terms and for their educational, psychological or legal implications (for example), but also as inflecting and reflecting—in a mutually constitutive way—more general relations of power, privilege and oppression. So I am formulating "child as method" as a way of reading the implication of modes of childhood within those broader socio-political relations. In my book I have explicated how this approach might work through analysis of the writings of the revolutionary anticolonial psychiatrist, Frantz Fanon, in particular tracing through the ways discourses of childhood contribute to form racism (and I specifically address this to the current situation of Brexit) (Burman 2018a, 2019a). Elsewhere I have attempted to illustrate its fruitfulness as a resource alongside but also offering a critical commentary on cultural historical and activity theory (Burman 2018b). I have applied to formulate a more democratic open pedagogical approach to the study of childhood using visual methods (Burman 2019b) and—with a colleague—to explore how the domain of nonformal, out of school educational contexts can illuminate the history and current practice of what we understand by schooling, and its limits (Burman and Miles 2018). I am fortunate to be working with some talented doctoral students who are now developing "child as method" to apply to practices of datafication in early childhood education (Pierlejewski 2018, 2019), to inform a cross-national study of early childhood educational practices in China, Hong Kong and Singapore (Zhou et al., in press), and to interpret current debates on educational practices around transgender children as offering a lens on contemporary social and educational understandings. Moreover, the newly emerging research on postsocialist childhoods (Silova et al. 2017, 2018) is developing this perspective further.

In recent times there has been much discussion about what is considered the medicalization and psychologization of childhood. In the last few decades, completely new diagnostic categories have emerged, such as ADHD, dysgraphia, dyscalculia, etc. In your opinion, are these tools based on normative assumptions that aim to

"trim down" individuals to the needs of our systems, or by making such diagnoses, are we rather discovering real problems and helping children in their development?

Here you are setting up an opposition that is—for better or worse—untenable. Yes, I think that these new(ish) childhood diagnoses may well work as ways of policing and regulating, constructing even, compliant appropriate childhoods. The rise (and rise) of ADHD under conditions of intensification of schooling and knowledge-based economies seems to support this. On the other hand, there is also no question but that these labels also produce the forms of experience they identify. Ian Hacking (drawing on Foucault) calls this the "looping effect," and describes this very well in terms of how we (children and adults) become invested in such descriptions (Hacking 1995). Hence the answer to your question is yes and yes, but neither should be seen as a good thing.

This opposition is not something I fabricated. It exists in the minds of my many parents, educators and teachers, people I often talk with. I'm afraid your answer "yes and yes" may not be very practical for them. Is there any practical way to break out of this cycle?

I think it is helpful to see how we come to identify with and invest in available forms of categorization in ways that become personally meaningful; that is, they subjectify us. Attending to this process at least gives the chance of being able to notice what is happening and so allow people to take some critical distance from these, or propose alternatives. The rise of the discourse of neurodiversity, for example, suggests that counterdiscourses to the mainstream medical approaches are flourishing.

Perhaps a not so common but nevertheless very serious problem is that of children in court as witnesses. What are your thoughts on this?

What about it? I don't see that children as witnesses pose significantly more problematic issues than adults as witnesses. Many of the same phenomena noted about children are just as present in adults (e.g., feeling under pressure to formulate answers to silly questions when they are posed by authority figures; interpreting being asked the same question twice as meaning that your first answer was wrong, both of which were criticisms levelled at Piagetian research long ago, but documented in adults too). But a key problem that this question highlights is the mutual relation of circular legitimation between psychology and the law: Normative conceptions formulated in one domain get taken up as unproblematic in the other, and so each domain—perceived as independent—confirms the authority of the other. This has been called "developmental psycho-legalism" (Cordero Arce 2012, 2015). Not only does this naturalize what is a very contingent body of psychological knowledge and understanding about children's competences, capacities and susceptibilities to influence (or compliance), but it also confirms the moral-political order as inscribing those developmental norms and trajectories.

Are there any solutions applied somewhere in the world you know of that would prevent or at least limit this psycho-legalism?

In fact it's the other way round. This is happening at the transnational level and contributes what amounts to the globalization of specific cultural-political norms and practices around children and childhood. It's there in UNICEF policies, in World Bank documents and so on. Limiting or preventing this is a matter of local resistance and critical and cross-disciplinary transnational solidarity work between academics and activists.

In subsequent editions of *Deconstructing Developmental Psychology*, you point to a number of issues concerning not only childhood, but also development processes in adults. Which of them do you consider to be the most urgently in need of solutions at the moment?

Poverty, racism, war, inequality, oppression, neocolonialism, displacement, impending climate catastrophe—shall I go on? I guess this is not the order of issue or problem that you meant, but really it is these that we need to attend to, in the sense of addressing how psychology and other disciplines of normalization and pathologization engage supportively with these issues instead of propping up those regimes and systems that contribute to and maintain them.

Here your views coincide with those of some of my other interlocutors, who also stress that psychology should focus more on issues that are central to our field. In your opinion, is it enough to change the subject of the research, or is it necessary to change the methods we currently use?

Clearly both, as well as to change who "we" are. That is, who gets to become psychologists and who comes to represent the discipline... In British universities, students across many disciplines are asking questions about "why are all my professors white?" and "why is my curriculum white?" Psychology has a problematic recent history as being seen as a feminized discipline (although as I note in my book *Deconstructing Developmental Psychology* it wasn't always like that!), and I suppose I have benefitted from becoming a teacher and researcher at a particular historical moment and in a particular geographical location which—after significant struggle—managed to be more receptive to feminist arguments. However, the absence of black and minority faces and voices in our departments speaks to the limits of the discipline and reproduces its exclusions. Clearly I am not just saying we need more diverse people to become psychologists—there's much more than that needed—but it would be a start.

Let's discuss some more general topics for a moment. For a few years now, there has been heated discussion about a crisis in psychology, and the term "replication" is eagerly used to describe it. I am convinced that this crisis is much more serious and goes far beyond the problem of the replicability of research results. But when reading the work of psychologists like yourself, who take a critical approach, one's eyes are opened to additional issues that don't even occur to someone educated in traditional academic psychology. In your opinion, what are the main problems facing modern psychology?

The "crisis" in psychology ushered in many fruitful critical debates and resources. I don't see "replication" as big problem, but as an impossible project indicating the omnipotent desires of psychologists. As social theorists (from Jacques Derrida and Judith Butler onwards) have been saying for some time, even repetition does not reproduce similarity. There is always difference, of time, of space. Re-iteration is not simply iteration. The quest for prediction and control engages with a pitifully reduced model of the world and phenomena that the criticism of lack of "ecological validity" scarcely names. Even seeking generalizability cannot stand as a criterion without much further elaboration, explicating what is being generalized, why and how—and once you do that all the complexity and indeterminacy of everyday material social life lived in specific historical geopolitical conditions flood in. As a reflexive discipline, subjectivity—or what positivists try to screen out or characterize as "bias" or "volunteer effects"—always insinuates inside the traditional measures. In my opinion, the more "objective" psychology tries to go—into neuroscience, for example—the more we see subjective experimenter and interpreter effects creeping back in unacknowledged. (But fortunately some commentators—both critical theorists and neuroscientists themselves—are on to this.) As feminist scientists, including philosophers and historians of science, have said for many years: There is no view from nowhere; all knowledge is situated and partial—partial in the double sense of being perspectival (and so necessarily aligned) but also thereby incomplete. Psychologists need to give up the project of grand universal claims to knowledge that not only prop up oppressive institutions (that require such models relying on normative developmental models, such as early intervention programs, etc.) but also on closer scrutiny fall flat on their faces because of logical incoherence.

What ways of solving these problems do critical psychology propose?

There isn't a unified critical psychology, so I guess there would be lots of proposals of ways to answer this question. Some critical psychologists would rework or apply existing psychological concepts and methods differently, to come up with different practices. Others would start from a different place, building new psychological concepts and tools from other—less oppressive or tainted (if that's possible)—cultural backgrounds to formulate different modes of partnership and activity. This is where some varieties of critical psychology share a background with critical social psychology and community psychology. Others would want to refuse the use of any psychological concepts or practices, while still others would focus on exploring the complex history of how psychology came to take the forms that it has. There has been valuable conceptual, methodological and practical work on all these fronts.

The way in which we psychologists present certain issues has an impact on social practices. For example, the hydraulic model of emotion, taken from psychoanalysis, which says that suppressed emotions must someday escape and somehow be resolved, is still

present in our culture and constitutes a kind of explanation, or sometimes even an excuse, for many violent acts. There are many more similar thought constructs that have migrated from science to everyday language and thinking. Some of them are very harmful and dangerous. Which of them would you consider the most salient and why?

Sorry, I don't accept or recognize that model of emotion as psychoanalytic at all. But yes all kinds of bowdlerized ideas get taken up in everyday and policy discourse. Much of my work has been focused on trying to problematize the apparently banal incontestable "truths" of development that inscribe not only textbooks of child development but also policies on national and international development. So much of this is seen as "commonsense," and that commonsense not only comes from the popularization of psychology into everyday life (although this also does happen, within increasingly psychologized cultures) but these commonsense assumptions (about gender, heterosexuality, culture, etc.) enter psychological models too.

Do you think that we should focus on the difficulty caused in psychological research by the presence of colloquial psychological concepts?

Yes! Because what it shows is that psychological research is necessarily, inevitably implicated in and structured by those everyday ideas and practices. They can't just be stripped away. Plus acknowledging this exposes the sham of peddling back to people, in the name of psychological expertise, what are in fact just dressed up versions of commonsense ideas they already know. David Ingleby said it so well, long ago: an originator of the notion of the "psy complex," pointed out in an early critique "Psychologists claim to be social engineers, but turn out to really maintenance men" (1972, p. 57).

Both critical and feminist psychology are often met with criticism, sometimes quite harsh. Looking at what you do from the other side of the barricade, is there any particular type of criticism aimed at you that you feel is particularly serious and justified?

Again, I don't see there as being a barricade; or if so then there must be many as there are so many debates and discussions. But, yes, there are of course criticisms that need to be taken seriously. A first one that I have long encountered is that such criticisms are all about being negative, and do not offer anything positive. Actually this is one that I can empathize with but don't really rate as a major problem as in my view critique has to stand irrespective of solving the problems identified. Indeed Foucault offered a key response here, highlighting how moving too quickly to solutions threatens to reproduce the same problem (Foucault 1977). (This is also the problem with utopian thinking in general.) It also arises from a fundamental misunderstanding that sees being critical as being negative. Closely following this criticism is the claim that critical psychologists work only at the level of theory and don't change practice. Actually, this is far from the case. Indeed it's largely the other way round. It is already existing practices that have inspired the critical theorizing, often as an act of solidarity or partnership work in which the critical psychologists learn from the practitioners. So, for example, the various projects that I have been involved with—whether on transnational dynamics of violence against women and children, or local projects on impacts of welfare benefit cuts—have inspired my conceptual and theoretical work rather than the other way round. I take very seriously claims of indirectly maintaining or legitimating oppressive knowledge structures. But this is a condition of life as critical academic working in a University in a global metropolitan center. One is always compromised and complicit. Claims of irrelevance don't bother me much—I'd rather be understood to do nothing much, than do loads of damage.

If psychologists finally cease to take the reproduction of mental terms in everyday life as evidence for their existence as causal entities, will psychology then turn into textual analysis?

As a materialist, I don't see problematizing models of causality as turning all psychology into textual analysis. We are way off being able to say anything about causes in psychology, I think. On the other hand, it depends on the model of textual analysis. I subscribe to a model of discourse that sees these as systems of meaning grounded in history, economics and geography; while the material world can only be apprehended through representational systems structured by those features. The question of what the proper domain of psychology is or could be outside the colonial, liberal bourgeois, heterosexist conditions that gave rise to it remains open.

In your opinion, is it possible to combine the post-structuralist and realistic approaches in psychology? What would such a post-deconstructive psychology look like?

I don't accept that opposition between post-structuralist and realist approaches. Your question seems to suffer from the presumption that reconstruction would be a good thing, which—as I indicated above—I cannot accept. Instead, I guess there will be new forms of critical psychology that will emerge in relation to, and hopefully to counter, the non-critical varieties. If past experience is anything to go by, much of these will come from other disciplines and practices.

In 2018, Helen Pluckrose, James Lindsay and Peter Boghossian exposed their large-scale hoax in an attempt to ridicule what they called "grievance studies" (News at a glance 2018)**. In their opinion, areas dealing with discriminated social groups have been consumed by intellectual "corruption" for some time now. It is sufficient to dress up absurd or even immoral hypotheses or data into political and ideological jargon fashionable in certain milieus in order for them to be printed in prestigious magazines, and consequently to influence public opinion and politicians. Pranksters? Whistle blowers warning us against the invasion of science by ideology? Or maybe hooligans who tried to tear down important ideas? What are your feelings about this incident and its consequences?**

Actually I do have some small bit of sympathy for the descriptor "grievance studies," even though it can work appallingly to delegitimize experiences of oppression and marginalization. But how does this come to operate in a zero-sum world that presumes either that all grievances are legitimate, or they aren't? Isn't "grievance" too reduced and psychologized a description that already denies objective, material conditions of oppression? Do we not all carry with us effects of injury that demand attention, to be grieved over if not to be a source of grievance? Even though (as in the discussion of Hacking above) I might be concerned about what the investment in "victim" or "aggrieved" identities might do, it's important not to blame the victims further. Rather, the issue is the cultural conditions that produce such invested identities, alongside a misconceived model of identity politics that threatens to delegitimize group mobilizations on the basis of oppression. But, you know, even if this was a hoax at the level of international journals, such phenomena smack to me of being very USA, and like the US domination of Anglophone psychology, and globalized further, its influence has spread. It may be that arenas dealing with discriminated groups have been corrupted by the bowdlerized discourse they have mobilized, but whose responsibility is this? As you would guess from my earlier responses, I don't see science and ideology as separable, and indeed would challenge the ideology of scientificity that underlies traditional psychology. So rather than eliding the issue and the arena, I'd see this particular prank as a commentary on the gullibility and market pressures on the ever accelerating publishing industry, rather than about this specific topic (albeit it is the latter which has gained it its publicity). It is deeply troubling and indicative of the right wing alliances at play if this is interpreted as a warrant to disregard work attending to experiences of oppression and efforts to challenge and redress these.

I have asked all my interviewees about the most significant achievements of our field. I am very curious about your response. Do you perceive them in the achievements of traditional academic psychology, or rather in what critical psychology has developed?

I can't say I identify with any one field. I'm as much at home (or homeless) in history, development studies, feminist studies, sexuality studies, postcolonial studies, education, childhood studies, critical psychoanalysis—including psychosocial studies, as in psychology. On the other hand, I absolutely defend my claim to be a psychologist, and speak as a psychologist. It is, after all my intellectual history and biographical disciplinary trajectory (which also, by the way, includes cognitive science). As a critical psychologist, I see psychology as having been very successful in getting itself allied with practices of power of all kinds. The project of critical psychology must be to expose and oppose this.

What achievements of critical psychology do we enjoy on a daily basis without even being aware of it?

Mainly critical (and uncritical) psychologists are beneficiaries of all kinds of other critical, transformative work undertaken by countless others that we are often unaware of. Since we started off talking about anti-psychiatry, I'd say one obvious example would be the way approaches to working with people who are sometimes described as psychotic, schizophrenic or having visual or auditory hallucinations are now transformed by the hearing voices movement. However, it is the fate—under existing conditions of power relations—that such interventions become amenable to recuperation. Hence, having won the argument that people can "recover" from such experiences, we now have paradigms of recovery that are normalized that demand that people return to work and be compliant, economically active neoliberal citizens. We should therefore expect that an intervention that, in one time or place, has been transformative, might in another become reconstructed as oppressive or unhelpful. As you can see, I don't subscribe to a linear narrative of (social as well as individual) progress.

What are the greatest challenges facing psychology in the twenty-first century?

Making itself redundant. It has been too successful as worming its way into practices of power. Whether and how it could be part of emancipatory practices (rather than simply appropriating and depoliticizing its language and approaches) is a different, big question.

And what challenges and questions have you set yourself?

Most recently, I have been revisiting the work of the revolutionary psychiatrist and social theorist Frantz Fanon, reading this as a corpus in which to explore how claims about children and childhood work in normative and counter normative ways. This is part of my project to formulate "child as method," an approach to thinking children and childhoods as a means to understand wider sociohistorical and geopolitical conditions. Rather than simply fitting children and childhood into wider social theory, then, the idea is to see how models of children and childhood have worked and continue to work to produce those conditions in constitutive ways. Overall, it's part of a project to foreground postcolonial and feminist theories within critical psychologies and educational studies. It's exciting to see colleagues and students run with these ideas and find applications and formulations that I wouldn't have dreamt of to do really interesting and valuable work. So right now, this seems fruitful.

At the beginning of your career you yourself chose to defy the crowd following the dominant tendencies of the time. Would you encourage young people to adopt a similar attitude?

I didn't see this as a choice. It was what I did. I suppose, following this line of thinking, I'd say that it's important not to do work you don't want to do, or don't believe in. (In that sense I remain a liberal humanist.) I remember seeing other colleagues, postgraduate researchers in my generational cohort, becoming dispirited and dropping out after being pressured to undertake work on topics or using methods that they were uninterested in or opposed to. In that sense, I was lucky to find myself in an environment where I was able to find my own reference points, formulate my own intellectual and practical projects, and define and forge relevant research communities, even if (or perhaps precisely because) that context was not particularly prestigious.

Who would you name as a model to follow for them just getting their careers off the ground?

I don't like naming individual people as this buys into an individualist model that underestimates the role of social conditions and relationships and overstates individual influence. In fact this was a key point for us on the editorial board of the journal *Feminism & Psychology*, when it was suggested we reprinted and celebrated the contribution of key figures or stars, and we realized that it would be more accurate to acknowledge the historical contribution of key articles of books, some of which were written jointly. I'll just add one more thing. Every professional should read Fanon's resignation letter as medical director of a psychiatric hospital to consider the conditions for, and limits to, ethical-political practice. He wrote: "The function of a social structure is to set up institutions that are traversed by a concern for humankind. A society that forces its members into desperate solutions is a non-viable society that needs replacing. The citizen's duty is to say so. No professional morality, no class solidarity no desire to refrain from washing the dirty laundry in public, can have a prior claim. No pseudo-national mystification finds grace when up against the demand to think" (Fanon 1956/2018, p. 345).

What are other issues we have not talked about yet but you would like to mention to our readers? Such as a message about

contemporary psychology, a big issue or important question about psychology?

Yes. There is a big movement afoot in universities concerned with decolonizing knowledge and practices. Psychology is sorely in need of decolonization. At all levels. There is the question of personnel, or what we—in psychology—call role models. Students are now asking "why are all my professors white?", "why is my curriculum white?" Psychology has a lot to answer for in this regard, and the lineup in this book, perhaps unsurprisingly, is no exception. All Anglophone, all bar one from the Global North, we are part of the problem. White, mainly men, creating knowledge and practices that—implicitly if not explicitly—reflect white, middle class, androcentric and heteronormative perspectives. Psychology, together with psychiatry, has long played a role as apologist for class and colonial domination, and gender plays a key role in the articulation of this too. There are, of course, many indigenous psychologists and psychologies, many critiques of the skewed and limited perspectives currently on offer as hegemonic, supposedly general and universal psychological models. Where are the minority voices, the perspectives from the Global South, from the non-Anglophone world (whose traditions you might know much better than me perhaps)? Some key names, such as Sylvia Wynter, are now being remembered (see, e.g., Wynter 2003), but it is not only a question of adding in the missing voices and perspectives but rather really thinking through and working through what their exclusion or devaluation means for the kinds of psychology we are familiar with and should be practicing.

A final question. While working on this book, I asked my readers to submit one question they would like to ask the most eminent leading psychologist, and I received thirty-one questions. Would you agree to answer one or two of them?

Yes.

Please choose one or two numbers from 1 to 30.

Ok 7 and 27…

7. How can we reduce potential harm in psychotherapy?

Good question! Not trying to be like psychiatrists for one thing. And not losing sight of the person in the consulting room, or in the complexity of our various models. On the other hand, the theory is a tool to think with, to offer material to think with and through so it's important to be consistent in one's approach (I do worry about so-called eclectic approaches). Personally, I was drawn to train as a group analyst because this way of working unavoidably brings questions of authority and democracy into the room as matters that have to work with, rather than assumed or swept under the carpet. Social inequalities are neither reducible to therapeutic ones, nor can they be relegated outside the therapeutic context. We all participate in oppressive practices by living in an unequal world; as therapists at the very least we cannot further pathologize people for (if I may use such terms) not being sane in an insane world.

27. The most ridiculous idea in the history of psychology which was taken seriously?

So many to choose from. The concept of "race," which psychologists played a big part in developing and legitimating. And the most dangerous.

Selected Readings

Banister, P., Burman, E., Parker, I., Taylor, M., & Tindall, C. (1994). *Qualitative methods in psychology: A research guide*. Milton Keynes: Open University Press.

Batsleer, J., Burman, E., Chantler, K., Pantling, K., Smailes, S., McIntosh, S., et al. (2002). *Domestic violence and minoritisation: Supporting women towards independence*. Manchester: Women's Studies Research Centre, MMU.

Burman, E. (Ed.). (1990). *Feminists and psychological practice*. London: Sage.

Burman, E. (Ed.). (1998). *Deconstructing feminist psychology*. London: Sage.

Burman, E. (2017a). *Deconstructing developmental psychology* (3rd ed.). London: Brunner-Routledge.

Burman, E. (2017b). Developmental psychology. In W. Stainton Rogers & C. Willig (Eds.), *Handbook of qualitative methods in psychology* (2nd ed., pp. 450–472). London: Sage.

Burman, E. (2019a). *Fanon, education, action: Child as method*. London: Routledge.

Burman, E. (2019b). Found childhood as a practice of child as method. *Children's Geographies*. Retrieved from https://doi.org/10.1080/14733285.2019.1566518.

Burman, E. (2020). *Developments: Child, image, nation* (2nd ed.). London: Brunner-Routledge.

Burman, E., Aitken, G., Alldred, A., Allwood, R., Billington, T., Goldberg, et al. (1996). *Psychology discourse practice: From regulation to resistance*. London: Taylor & Francis.

Burman, E., Alldred, P., Bewley, C., Goldberg, B., Heenan, C., Marks, D., et al. (1995). *Challenging women: Psychology's exclusions, feminist possibilities*. Buckingham: Open University Press.

Burman, E., & Parker, I. (Eds.). (1993). *Discourse analytic research: Repertoires and readings of texts in action*. London and New York: Routledge.

Burman, E., & Parker, I. (Eds.). (1993, reissued 2013). *Discourse analytic research: Repertoires and readings of texts in action*. London and New York: Routledge.

Burman, E., Smailes, S., & Chantler, K. (2004). Culture' as a barrier to domestic violence services for minoritised women. *Critical Social Policy, 24*(3), 358–384.

Chantler, K., Burman, E., Batsleer, J., & Bashir, C. (2001). *Attempted suicide and self harm—South Asian women*. Manchester: Women's Studies Research Centre, MMU.

Levett, A., Kottler, A., Burman, E. & Parker, I. (Eds). (1997). *Culture, power and difference: Discourse analysis in South Africa*. London: Zed Books/Cape Town: UCT Press.

References

Burman, E. (2018a). Brexit, 'child as method', and the pedagogy of failure: How discourses of childhood structure the resistance of racist discourse to analysis. *Review of Education, Pedagogy, and Cultural Studies, 40*(2), 119–143.

Burman, E. (2018b). Child as method: Anticolonial implications for educational research. *International Studies in the Sociology of Education*. Retrieved from http://dx.doi.org/10.1080/09620214.2017.1412266.

Burman, E., & Miles, S. (2018). Deconstructing supplementary education: From the pedagogy of the supplement to the unsettling of the mainstream, *Educational Review*. Retrieved from https://www.tandfonline.com/eprint/iTwP73UzKHg8U8HfUf8C/full.

Cahalan, S. (2019). *The great pretender: The undercover mission that changed our understanding of madness*. New York, NY: Grand Central Publishing.

Chen, K. H. (2010). *Asia as method: Toward deimperialization*. Durham: Duke University Press.

Cordero Arce, M. (2012). Towards an emancipatory discourse of children's rights. *The International Journal of Children's Rights, 20*(3), 365–421.

Cordero Arce, M. (2015). *Maturing* children's rights theory. *The International Journal of Children's Rights, 23*(2), 283–331.

Fanon, F. (1956/2018). Letter to the resident minister. In F. Fanon (Ed.), *Alienation and freedom* (p. 345). London: Bloomsbury Academic.

Foucault, M. (1977). Revolutionary action: 'Until now'. In M. Foucault (Ed.), *Language, counter-memory, practice* (pp. 221–233). New York: Cornell University Press.

Hacking, I. (1995). The looping effects of human kinds. In D. Sperber, D. Premack, & A. J. Premack (Eds.), *Symposia of the Fyssen Foundation. Causal cognition: A multidisciplinary debate* (pp. 351–394). New York: Clarendon Press/Oxford University Press.

Ingleby, D. (1972). Ideology and the human sciences: Some comments on the role of reification in psychology and psychiatry. In T. Pateman (Ed.), *Countercourse: A handbook for course criticism* (pp. 51–81). Harmondsworth, UK: Penguin.

Mezzadra, S., & Neilson, B. (2013). *Border as method, or, the multiplication of labor*. Durham: Duke University Press.

News at a glance. (2018). *Science, 362*(6411), 134–136.

Pierlejewski, M. (2018). Constructing deficit data doppelgängers: The impact of datafication on children with English as an additional language. *Contemporary Issues in Early Childhood*. https://doi.org/10.1177/1463949119838089.

Pierlejewski, M. (2019). The data-doppelganger and the cyborg-self: Theorising the datafication of education. *Pedagogy, Culture & Society*. Retrieved from https://doi.org/10.1080/14681366.2019.1653357.

Romme, M. A. J., & Escher, A. D. M. A. C. (1989). Hearing voices. *Schizophrenia Bulletin, 15*(2), 209–216.

Silova, I., Millei, Z., & Piattoeva, N. (2017). Interrupting the coloniality of knowledge production in comparative education: Postsocialist and postcolonial dialogues after the cold war. *Comparative Education Review, 61*(S1), S74–S102.

Silova, I., Piattoeva, N., & Millei, Z. (Eds.). (2018). *Childhood and schooling in (post) socialist societies: Memories of everyday life*. Basingstoke: Palgrave.

Wynter, S. (2003). Unsettling the coloniality of being/power/truth/freedom: Towards the human, after man, its overrepresentation—An argument. *CR: The New Centennial Review*, *3*(3), 257–337.

Zhou, L., Burman, E., & Miles, S. (in press). Teachers' perspectives on 'learning through play' in Mainland China, Hong Kong and Singapore kindergartens. In M. Salazar Pérez (Ed.), *Sage handbook of global childhoods.* Thousand Oaks: Sage.

13

Brian A. Nosek: Open Science and Reproducibility Projects

It's hard not to see the misalignment between what is good for me as a practicing researcher and what is good for science.

Brian A. Nosek

T. Witkowski, *Shaping Psychology*,
https://doi.org/10.1007/978-3-030-50003-0_13

Like some of my other interlocutors, Brian Nosek became a psychologist by accident. When he entered the psychology field, he was a computer engineering undergraduate. Toward the end of his third year he enrolled in psychology classes as a break from the really hard courses. He soon realized that he was actually spending all of his time thinking about and working on his psychology courses. He found it much more interesting to do science on humans than to do research on circuits. When he came to the conclusion that he could really make a career of doing science on humans, he was hooked. The following year he switched his major to psychology but finished his degree in computer science as well.

When pursuing his graduate degree in experimental psychology, he started working on the Implicit Association Test, which reveals people's implicit prejudices with the push of a button. Tap right every time a male name appears on a screen, for example, and left for a female name. This seemingly easy task gives interesting results when you add to the list of names some stereotypically related roles. Even the most liberal minds will sometimes stall when asked to press the same button for the word "executive" and for the name "Susan." The tests are challenging, informative and kind of fun. So, in 1998, Nosek convinced his mentors, who had developed the test, to put it online. He called it "Project Implicit." It was a success: About a million people per year now take the test for research, corporate training and other applications. It has done much to spread knowledge about the nature of implicit bias.

Although Nosek became a psychologist by accident, his approach to his work is certainly not accidental. As he says himself, his current activity is the culmination of his lifelong experiences. Both his mother and father, in different ways, lead their lives according to the values which are important to them. His father is a manager for whom ethics and honesty have always formed the basis of people management. His mother worked for the church, providing religious education, and was always focused on that which would improve her work.

Apart from his devotion to the values he took from home, another factor which has helped him in his projects has been his "ignorance." As a graduate he was totally unaware of just how far scholars are resistant to innovation. In his opinion, despite the fact that there is still much said about innovation, especially in methodology, there is a strong tendency to accept a system which functions. Nosek considers that if, as a graduate, he had been aware of how strongly the status quo was accepted he would never have undertaken any of his projects, either those which he completed successfully or those in which he is currently engaged. Many times, when his applications for research grants were being reviewed, he was called "Pollyanna" because of his optimistic approach to innovative research methods and tools. However, he also remembers lessons from his mentors, who asserted that not attempting to do something was a guarantee of failure.

His ethical attitude learnt at home, incorrigible and unadulterated optimism through bitter experience, as well as his experience gained through cooperation with many scientific institutions throughout the world during the coordination of Project Implicit in 2011 enabled him and his collaborators set up the Reproducibility Project, with the aim of trying to replicate the results of 100 psychological experiments published in respected journals in 2008. During the Reproducibility Project, in 2013, Nosek took leave from his post at the University of Virginia in Charlottesville to co-found and direct the Center for Open Science (COS), a non-profit company that builds tools to facilitate better research methodology. COS continued coordinating the Reproducibility Project, whose results were published in 2015 in *Science*, and found that only 36 of the 100 replications showed statistically significant results, compared with 97 of the 100 original experiments.

COS has built a team that today numbers around 50 people from across disciplines, such as astronomers, biologists, chemists, computer scientists, education researchers, engineers, neuroscientists and psychologists. COS operates the Open Science Framework (OSF), a special online service where scientists can preregister their research, document their research process and share their data, materials and project outcomes. While OSF initially focused on psychology, it has since broadened to encompass any research field. COS simultaneously runs a number of research projects, including Many Labs I, II, and III, Reproducibility Project: Cancer Biology and many more.

Nor has there been a lack of inspiration and motivation drawn from Nosek's private life, and one experience had a particularly strong influence on his engagement in the open science movement. In the spring of 2011 Sarah Mackenzie, who had been a witness at Nosek's wedding, was diagnosed with a rare form of cervical cancer. Sarah and her family were strongly motivated to discover as much as possible about the disease and assure her of the best of care. They were not scientists, but they started to go through the literature in search of suitable articles. One evening a very angry Sarah called to Nosek, complaining that every time she found an article which might be of significance in understanding her disease, she came up against a demand for the payment of 15–40 dollars for access to the article. Research which had been financed by public money was inaccessible to her unless she paid. Nosek gave Sarah access to his subscription account so that she could comfortably peruse the scientific literature, but he woke up to the fact that the majority of people in Sarah's position did not have the luxury of having friends working in wealthy academic institutions which were in a position to pay subscriptions to most of the scientific periodicals. Such a situation demanded change.

In 2015, Nosek co-authored a set of guidelines for transparency and openness that more than 500 journals have signed up to (Nosek et al. 2015). That same year he was named one of Nature's 10 and was put on the Chronicle for Higher Education Influence list. As *Nature* wrote about him (365 days 2015), he is a "bias blaster" who has "pledged to improve reproducibility in science."

Prof. Nosek, you're one of the world's most well-known advocates of the open science movement, engaged in improving transparency in science. What led you go beyond your research on implicit cognition and engage to such a significant degree in improving research methods and collaboration between scientists?

My interest in these issues started in graduate school, when I took Alan Kazdin's research methods course. This was around 1997, and he had us reading papers from the 1960s and 1970s by Bob Rosenthal, Tony Greenwald, Jacob Cohen and Paul Meehl, where they articulate challenges like publication bias, lack of replication or low power, and they outlined solutions. For example, let's try and do more replications, let's increase the power of studies. Even preregistration was mentioned in some of these articles. It was shocking as a grad student in the late 90s to think that methodology has been outlining these problems and solutions for 30 years, but nothing has changed. Why is that? So, like many graduate student cohorts, we would have discussions after lectures or meetings at the bar and talked about how we would change the system if we could. And, of course, we didn't have the opportunity to make any substantial changes at the time, but the interest in these issues remained at the core of how I thought about the research we did in our laboratory.

So we started with trying to address the power of our designs. We created a website for collecting data about implicit biases, which became very popular. We achieved very high power tests of the kinds of questions we were investigating. When I became a faculty member, we made it a routine practice to share as much of our materials and data as we could on our personal websites. And when services came up, we tried to adopt things that would improve our sharing. In the mid-2000s I started to write grants to the National Institutes of Health and National Science Foundation to create what we called at that time an open source science framework. We had a technical lab for a long time, we operated this website for collecting data, and we thought it would be useful to have a service that would make it easier to share that data and for others to use it as well. But we couldn't get it funded then because there was just a wide range of reviews. Some said this was very important, a necessary change, and others said that people don't like sharing data. It just wasn't the right time. But there was a general interest that we had as a laboratory in improving the process of our own work, building tools for the technical portion of what we do, to make it easier for others to do it.

Then, in 2011, a lot of these methodology issues became of broad interest to the research community because of the Diederik Stapel fraud and because of very surprising results being published in leading journals, like Daryl Bem's ESP work. And then paper in *Psychological Science*, "False-Positive Psychology" (Simmons et al. 2011), really crystallized for many people how we have some practices whose implications for the reliability of our results we don't really understand. They provided a rhetorical *tour de force* of how that happened, which helped people to understand those behaviors and their consequences.

So all of that is happening around the same time. One of our failures is with replication. Initial studies became public, and people failed to replicate them. As a laboratory, we had been thinking and talking about a lot of these issues for a long time, but we didn't have anything concrete, no database of evidence of replicability of findings. We had individual studies that we failed to replicate, but that happens all the time, and you can't really tell if there's anything systematic. We decided to think about how to start a collaborative project, to see if we could replicate some meaningful sample studies in the field. So we have this replication project that turned into the Reproducibility Project: Psychology. We've just started sharing it with others to gauge interested. Lots of people got interested very quickly, and that became a big project.

Here's the background on starting the COS: Jeff Spies was a grad student in the laboratory with a history as a software developer, and we were looking for ideas for his dissertation. We kept coming back to this old project idea of an open source science framework. And while it was unusual, he decided it was very much in line with his long-standing interests, so he started working on that as a dissertation project. So we had this replication project and this technical project that received some broad attention, which we conceived of as lab projects. We were self-funded, without grants or anything, but then they came to the public's attention, and some funders started calling. Very rapidly we moved from interesting but small projects to thinking about doing things in a big way. Within a matter of two months we went from a small lab to launching the COS as a non-profit.

Your activity has brought you global recognition, but also a lot of enemies. I am thinking particularly of speeches like Norbert Schwarz, who said that replication is important, but it's often just an attack, a vigilante exercise. Still others have described you as replication bullies, false-positive police and data detectives. Susan Fiske called people who publicly discuss the results of scientific research and the methodology, methodological theorists. Did you expect such harsh attacks, and how do you deal with them?

I think I both expected and did not expect that kind of criticism. Of course, there will be criticism when moving into this kind of work. And that isn't a bad thing. A lot of what this movement is about is challenging some of the fundamentals of how we do science and training a skeptical eye on the credibility of our existing evidence space. If that didn't receive a lot of pushback, it would be very surprising. That would mean that we are all putting on a very cynical facade and don't actually believe in the research that we do. Of course, there is going to be vigorous debate and pushback over that. I did hope that it would remain more scholarly and collegial than it has been at times, but this is getting a lot to people's core identities. So the fact is that people respond very strongly, emotionally, with a real sense that this is an attack not just on the science, but on me and my colleagues. It's not unreasonable that people have those reactions.

It's important to keep the message very clear regarding what this is and isn't about—the replication movement that we have tried to be involved in, or about our core principle of always remaining self-skeptical. I think that is a critical value of science, that we should be the most skeptical of our own results. The public at large depends on us to be critical of our own work, so that whatever comes out at the end is something the community can rely on without enjoying the expertise that we have at the beginning. It isn't about trying to say someone is a bad researcher or doesn't deserve to be a part of the scientific community. Those are very personal attacks. And a lot of that does occur, because people, whatever their point of view, sometimes can't help themselves. Sometimes they do dislike a person or an idea, but we try to be a moderating voice in that aspect, always with respect to the individuals and gratitude for positive engagement. It's about rescuing the science and always trying to be ourselves, living up to the values that we espouse: transparency, reproducibility, openness to criticism. The more we live those values and put them on display, I think that has, to some degree, helped to moderate some, although not all, of that criticism.

Open science can be seen as a continuation of, rather than a revolution in, practices begun in the seventeenth century with the advent of the academic journal, when the societal demand for access to scientific knowledge reached a point at which it became necessary for groups of scientists to share resources. The well-known sociologist of science Robert Merton mentioned communitarianism as one of the

foundational values of the scientific ethos. Where did such ambivalent attitudes toward the open science movement come from among those scientists who are the heirs to these traditions and values?

That's a great question, because at the core of what we're pursuing as a culture change enterprise is shifting incentives, norms, and policies for science to be closer to those Mertonian norms. We can easily recognize them as the core values of how science is supposed to operate. How we got away from that is a complex question; even Merton himself didn't think that individual scientists necessarily operate by those norms. His focus was rather on how social structures, the system, needed to live up to those norms, even if individuals themselves were more focused on their own interests, secretive of how they were doing their work, that sort of thing. But the system needed to operate with that self-skepticism as a system.

That said, we can do better, we can live up to those norms even in our own individual behaviors. And I think that one of the things preventing us from living up to that is at a very base level, which is technical. It has never been as easy to share methodology and data openly as it is now. The internet has changed what it means to be open and sharing. My colleague Tony Greenwald was editor of the Journal of Personality and Social Psychology in the mid-70s, and throughout his editorship he required authors to physically send their data to him for analysis. That kind of expectation in the seventies wasn't very feasible. Now, it is trivial. So that is a change bringing us closer to the Mertonian norms.

A second factor, in my view, is science as a competitive system. It's hard not to see the misalignment between what is good for me as a practicing researcher and what is good for science. The policies, the norms, and the incentives that we create for how science is now rewarded, and how those structures work, have not been designed to facilitate the norms as it could. That is a key goal for us as an organization: altering policies, norms, and incentives so that researchers can live by their values while reinforcing scientific values at the same time. They can succeed and do science in a wise way rather than having those being in conflict.

The profession of scientist is characterized by almost unlimited freedom. They choose their own subject and methods of research,

That is a great question and I would love a good historical analysis to determine which of many possible reasons is the right one. In one sense, we've known about so many of these issues for a long time. The kindling for the reform movement has been there, and it has been accumulating for many years, so it had to happen sometime. Let's say 2011 is the starting point. It could be an accident of history that this was the year it finally started. Another possibility is that all that was accumulating, and some stimulating, singular events occurred that made it a lot easier to confront this at the scale we're doing now. The Stapel case, the Bem's paper, the Simmons, Nelson, Simonsohn "False-Positive Psychology" paper. These individual cases captured a great deal of attention beyond just people who care about methodology, where this conversation has been ongoing since the 1960s. Those provided some stimulus to really ask "what are we going to do about this?" or "what is this, really?" The fact that they happened close together in time also may be a factor. Instead of just one event, they all started to pile on, and it's like a dam broke.

Another factor is that this issue is salient not only in psychology. It has been accumulating attention across other fields. Maybe, with the internet and social media, rather than just the thought that in my little field we have this problem, it is a lot easier to see outside our disciplinary silos and say, "my goodness, biomedicine, they're having this problem too. Oh, economists, they're having this problem. Hang on a second. Who isn't having this problem?" Connecting those communities of people who care about these issues across disciplines may have facilitated collective action that made it a lot easier for the movement to become better, faster, and more impactful. I suspect that it's a complicated mix of many causes, but it's fascinating that even though those things have been known for a long time, these events congealed into a real movement this time.

Observing the behavior of scientists, I have also noticed that some of them, despite all the problems we are talking about, are full of optimism and announce to the world that we are in an age of great replication which will solve all the problems, and we are entering the straight and broad path to the truth. Meanwhile, despite some grounds for optimism, the reality looks a little less optimistic. In the PsycARTICLES database I checked exclusively peer-reviewed

articles. In 2017, 5166 of them were published, of which only 14 contained the word "replication" in the title. In 2018, their total number was 5530, of which 12 were replications; in 2019, 6444 were published and 97 contained the word "replication" in the title. In total, articles on replication accounted for around 0.7% of all publications. Are you yourself one of those optimists who believe we are at the threshold of a better tomorrow in our field?

Yes. I am optimistic. And the reason I am is that a sustainable culture change is growing. But this doesn't mean it's happening fast. There are a lot of people in this movement that are very discouraged. They think, "we've worked this out already, why are people still doing it wrong? Let's just get it done." But culture change is hard; changing people's behavior is hard. We have great literature that shows how hard it is to make these changes. Let's pay attention to the parts of our literature that we can trust and apply it as effectively as we can to the movement as it is occurring. There is good reason to be optimistic, because the core challenges are becoming very well known as problems, and it's hard to be in our field and not be aware of these challenges. That's a big first step. Whether people will change their behavior or not remains to be seen.

The second reason for optimism is that training is changing. Methodologists care about these things fundamentally and many are changing and updating statistics and methods courses. That matters, because training will stick with the generations as they come through this gradually.

The third reason for optimism is that the stakeholders, founders, publishers, societies, and institutions are paying attention. Maybe not quickly, but they're all changing their policies, their norms, and their incentives. Policy change is the best way to have sustainable change in the long term. The general shift from not requiring any transparency to encouraging or requiring openness fundamentally changes what comes through a journal. Incentives for preregistration and badges that make visible that other people are doing these behaviors, can have long-term accumulated consequences of forming new norms. Finally, Registered Reports are now offered by more than 200 journals. Registered Reports eliminates publication bias by conducting review and committing to publish results before the outcomes are observed. That is a fundamental change to the publication workflow and will have a lasting impact if it achieves broad adoption.

All of the critiques of the movement saying that it's not changing very fast are correct, but I don't know if it can change faster than it is. The groundwork is being laid. The changes we've seen are not superficial. It's not just one researcher and one team did something, and tried to replicate some findings, and that's the end of it. Changes are happening in the structure of how we do our work, and that's really what will help sustain it the long term.

The seriousness of the crisis in psychology is often diminished by describing it as replication crisis. We know, however, that it consists of many more problems than just the lack of replication. Which of them can be solved by the open science movement, and which should be overcome some other way?

Replication is the low-hanging fruit of improving research practices—here is a finding, here is a methodology, let's try to see how reliable that methodology is. From this point of view, the replication part of the movement helps stimulate attention to the broader issues of how we can make research progress as quickly as possible. Replication doesn't solve problems like construct validity, attentiveness to how our theories get refined and formalized so that they are testable, or connecting our theories with our operationalizations and inferential tests. There are big challenges in how we reason, how we accumulate evidence, and how we combine that evidence into theories in science. Open science doesn't solve these, but open science is an enabler of pursuing solutions to these challenges.

We have a lot to mature in how we develop theories and how we make connections. But we can't that work without open evidence, without transparency on how we're getting our claims, without better sharing, without materials being open and accessible, and without some replication to test our theoretical positions in a more formalized way. To me, the movement, at least in psychology, is finishing phase one, where the major theme was replication. I think it's entering or midway into phase two, which is about generalizability. It's not just about replicating a finding in that context, but about the breadth of where we can see that finding across different operationalization, across different samples, across different conditions of the experimental context. Next, I think, it's going to be entering a phase focused on measurement and theory. Now we see what is reproducible on our base claims, how we can better connect this evidence to more theoretical claims?

I see one more problem plaguing contemporary psychology, and which probably cannot be solved by the open science movement, or perhaps even deepened. I mean the shift from direct observation of behavior, widely regarded as an advance in the development of scientific methodology, to introspection. This was demonstrated in an outstanding 2006 article by Baumeister et al., and recently confirmed by Doliński (2018), who replicated Baumeister's investigations. Both articles show that over the last few decades, studies of behavior have become a rarity among psychologists. This issue is brought into sharper relief by the fact that the first of the two articles was published in the middle of the decade that the APA announced with great pomp as the decade of behavior. What are your thoughts on this issue?

This is interesting, and I agree that it can't be different than how we think about what the open science movement is trying to solve. We have to consider the incentives that shape individual researchers' behaviors. If the focus of my attention in research is to publish as frequently as I can, in the most prestigious outlets that I can, then I have to make practical decisions about the kinds of questions that I investigate. Measuring real behavior, that's hard. Sending out a survey, that's easy. I can generate more publishable units doing that.

Or using Mechanical Turk.

Exactly. Researchers are already stressed about publication demands, and now open science movement is increasing the pressure for higher power. This makes it even harder to do those behavior studies—now, instead of 50 people you need 500. This is a real issue, elements of the open science movement—increasing power, increasing transparency, increasing rigor—could be at odds with the goal of getting the research community less focused on self-report surveys and more diverse in how it measures human behavior. If we don't solve the incentives problem, then we will become a very narrow discipline. In part, that may reflect how hard is to study the things that we study. We have to recognize where we are as a discipline in terms of our tools and instrumentation, to be able to do some kinds of science. Then, we have to be realistic about what science can be done effectively with the available resources.

In some ways, we have tried to study questions that we just don't have the technology, the ability, or the power to study effectively. If we take seriously what questions can we productively investigate with the resources we have available, then we may recognize that a lot of the questions we want to study, cannot be studied effectively the way that we do science now. If those questions are really important, then we need to change how we do things. This is where another element of the open science movement has part of the solution–collaboration.

Right now, we are rewarded for being a vertically integrated system. Myself or my small team comes up with an idea, designs a study, collects the data, writes the report, all on our own. This requires lots of resources devoted to small groups. If we move to more horizontal distribution, where many different people can contribute to a project we can study many questions that we are presently not able to study well. There might be one team designing the study, and 15 teams that contribute data. With that kind of model, if we get the reward system worked out we could study some questions more productively and with adequate power to actually make progress on the problem.

The reform movement needs to attend to the effects of each change on the complex system of incentives and reports: What happens with increasing expectations for power? What will that change in what people study? How do we solve that? These systems are complex and getting all the incentives aligned for maximizing progress is hard.

There is another problem that is troubling psychology, and which you very diplomatically described in one interview as "conceptual redundancy." We know that it is basically about cluttering our field with needless, often duplicate theoretical constructs, about unnecessarily publishing and creating new concepts for previously known and described phenomena. This conceptual redundancy is increasing at a rather alarming rate. What is your opinion about it?

I think it's another illustration of a different part of the reward system. In psychology we value the idea that each person has their theory, and their conceptual domain that they study that is linked to them and their identity. The consequence is that this incentivizes the splitting approach, where the same concept is being studied by five different groups, each giving it different labels. Early in research, this can be useful. When you're in a generative phase, we have no idea what this problem space is like, and we need various approaches to explore that problem space. What if we don't have is the consolidation phase, where we say, "OK, there are seven different groups that have what seems to be the same kind of idea." How do we figure out what actually is the same and where the differences are? That lack of consolidation leads to a very fractured and not very accumulative discipline.

The social challenge we need to address is that we're individually tied to the words we use to describe psychological phenomena—I don't want my perspective on self-enhancement to be combined now with somebody else's idea that is the same, but uses a different word to describe it. The methodological challenge is to create occasions for similar ideas to be confronted against one another. And there is a great example, in development now, that I think would be useful as a prototype, and that I am hoping will become ordinary practice. The Templeton World Charity Foundation organized group of neuroscientists who all have different theories of consciousness. The scientists are told "you're going to sit together in a room for three days, and you're going to come up with experiments where at least two theories have different expectations of the outcome." And of course, it took them two days of yelling at each other to even figure out what the differences in their theoretical expectations were. But if we can stay scholarly, this process can be very productive to provide some clarity about the actual similarities and differences between these theoretical perspectives. They did come up with two experiments for which the theories make different predictions. And now they have funding to do the experiments. That is a great exercise, so I am hoping that some initial prototypes of that will be very widely disseminated, and generate lots of attention for how creating a confrontation and consolidation process to counterweight the spreading process as you introduced.

Daniel Kahneman proposed something that he called the adversarial method, but in a database of scientific articles I was able to locate just a few empirical works carried out in accordance with the recommendations of that method, and a few dozen publications involving discussion of adversarial collaboration.

Yes, it is hard to do. There are not many obvious rewards for entering into adversarial collaboration; people prefer to think "my findings will stay my findings, let's just ignore each other, we'll both be fine." With confrontation comes risk, so there is a social barrier to adopting it. Even though people might recognize that the value to science is high.

In your opinion, is there any chance in the near future of a unifying theory in psychology that would reduce this redundancy a little bit?

I'm skeptical, at least for the near future, because the phenomena that we study are so diverse that there probably isn't a unifying theory that would be tractable and useful. I do like the emergence of theoretical frameworks, like evolutionary psychology, for example, to organize a set of ideas about human behavior. But I suspect that we are going to have a few different theoretical frameworks for a while. I can't predict what convergence is possible.

Critics of replication projects have sometimes stressed that a strong emphasis on replication may lead scientists to focus on replication research instead of exploring difficult and serious problems that can have a significant impact on our reality. It is difficult not to admit that they are right. How do you resolve this dilemma?

I think this is misspecified critique. The idea that we don't replicate and we'll make progress without doing replication, is, I think, why we facing the problem we have. We have failed to understand the value replication for theory and discovery. We generate lots of ideas, we have the feeling of making progress, and then we have little actual confidence or credibility in the underlying basis of the claims that we are making. The fact that there are hundreds and hundreds of studies about ego depletion, and we're still debating whether ego depletion actually occurs, shows how problematic it is to not do things that increase the credibility of the core evidence.

The other thing that I think is misspecified about the argument is that replication is a mechanism for theoretical advancement. Replication is sometimes better for theoretical advancement than pure exploration. To do a replication, one must construct a situation in which you have a clear prediction of what should occur based on the prior findings and their theoretical context. Replication is the confrontation of an existing understanding of a phenomenon. The consequences of the outcomes of a replication are to affirm and create more generalizability for the existing understanding, or to confront that existing understanding with disconfirming evidence. The latter is inevitably theoretically generative because our expectations from prior findings were violated. The combination of replication and explication is one of the most theoretically generative things we can do. I don't know what the right ratio is between them for maximizing progress, but I think it is incorrect to say that replication is itself different than making theoretical progress.

Is there any particular type of criticism aimed at the open science movement that you feel is particularly serious and justified?

I think it's all serious and justified, unless it's personal. The claims of the open science movement need to survive critique, just like all of the findings that are being critiqued in replications need to survive their critique. Let me give a couple of examples. There is a critique that preregistration will reduce creativity, make research more boring. I don't have any evidence to show you that it will not do that. And we can generate plausible stories for how it could. For example, if I have to pre-commit, I might get more conservative, because pre-committing to crazy ideas might be embarrassing, particularly if they are wrong—which most will be—they are crazy ideas after all. But, some crazy ideas are totally transformative. Preregistration could induce risk aversion. If I don't have to pre-commit and I can just do it, then maybe I'd be more willing to take risks because who cares if I am wrong.

Whenever we're pursuing culture change there is potential for unintended consequences. In fact, unintended consequences are functionally inevitable, because we don't know all the consequences of our actions. If the open science movement did not have a skeptical audience constantly evaluating what happens when we make these changes, we will end up doing some things that are counterproductive for research progress.

So, I am very glad that there is positive engagement of skepticism for these changes. What I don't like is when it gets personal, like the "methodological terrorists" kind of remark. This is completely unproductive. It is also unproductive when the skepticism is so strong that people don't even want to try. The whole purpose of research is to try something and see what happens, and if we are so ideological that we say, "I cannot share the data because it will screw things up," so I even don't try to share, or "pre-registration isn't even worth trying, because I am sure that it will reduce creativity," that isn't really engaging in research. Competing perspectives are valuable as long as we are studying it, learning something, and then figuring out how to do it better.

How is it that in spite of information appearing about the crisis and the general decline in trust toward psychology, the field seems to be attracting ever greater numbers of people? As a major, it is shattering records among students in both Europe and America.

It's attracting more involvement. My guess is that early students at the undergraduate level are, by and large, unaware of these issues. I think that the emergence of awareness—and this is speculation—comes later in the undergraduate stage at the earliest, which is perhaps after the commitment has already been made.

A second speculation is that a lot of people early in their career find the reform movement exciting rather than unnerving. I think many people are entering the discipline with the feeling that the field is emerging rather than that the field is in tatters, in crisis, or in decline. The reform movement itself is very healthy and expanding, so I think that is having salutary effects on engagement.

Some people claim that psychology attracts people who like fuzzy disciplines, where there are no precise claims and hard knowledge. There is a lot of room for interpretation and no responsibilities required. What's your opinion?

I think it's possible that there are people who like the generative and open-ended explanations for why things happen the way they do, and psychology allows some of that freedom. That may even impact some of the issues we have discussed the research process. Maybe it's harder to become a physicist and think "let's just explore the world as it is, I will generate my own theory of physics." The frameworks for understanding in physics are stronger perhaps limiting entry of those who want to be very exploratory and generative. But of course, when you're out on the edges of physics it is still very exploratory. All kinds of crazy ideas that come up in theoretical physics these days, it's an amazing, generative, and creative field. There is probably a range of motivations and interests that get people into psychology. If there weren't people who get more singularly focused and want to solve this or that particular problem, it would be problematic for psychology. Psychology has enough to study that an ecumenical approach to who gets involved in research and their approach to research is ok and healthy at the current stage.

Fortunately, apart from problems, our field also has a lot of achievements. Which of the existing psychological discoveries do you consider to be the most significant breakthrough?

I can't say confidently, because I'd have to review all of the literature to say which of those things are the most important. But what comes to mind as you ask the question is the astonishing progress that has been made on understanding visual perception from what we understood about how the visual system works in 1900. This is a massive transformation. We've learned so much, and so much of that work has been applied to computer vision and other kinds of research applications, and to our practical understanding of the visual system in animals and in humans. I am a total outsider to it, but I love what I know of the work.

Kahneman and Tversky's heuristics and biases is, I think, the most directly impactful on questions in the social cognitive domain of judgment. That's an obvious one to mention, but the degree to which we now understand motivated reasoning in the big picture and particular biases that are current reasoning is incredibly important. What we need from that field is to grasp how we deal with those biases in everyday judgment and decision making. We need systems and solutions to address unwanted biases where they occur. It would be transformative for human behavior if we can solve these questions.

The last theory would be areas where we have effective treatments for some areas of mental health. The fact that we can address many types of phobias, in a single session, or in 8 hours with cognitive-behavioral treatments, is astonishing.

Each of these examples shows that basic psychological knowledge can be translated into an understanding of how we improve human behavior.

What do you consider to be the biggest challenges in the field of psychological science generally?

I think the biggest challenge that we're spending our time worrying about is how to fix the reward system. This is not about research topics, but I think it has direct implications for research topics. And the part that is the hardest to change is hiring and promotion at academic institutions. If we don't fix the need for more and more publications and more and more prestigious outlets as the criterion, none of the other open science changes will be completely effective. It is very challenging because institutions have their independent, ad hoc criteria. There is no singular policy or decision-making body to facilitate this change. The reason I focus on systemic issues like that is that if we fix these issues, all the other challenges in the field of what research gets done and how it gets done will be much easier to solve. The systemic challenges of how people are rewarded for the kind of work they do are barriers to the more specific challenges of how to do the best research.

What are your thoughts about artificial intelligence and about the future of artificial intelligence in psychological research?

There is a lot of potential in what artificial intelligence approaches can provide as a complement or as a replacement for some of the things that we're doing in actual research process. It's exciting to see the advances being made, including the identification of serious challenges, the black box problem being the most obvious one. If a machine solves the problem, but we have no idea how, then what have we learned, and how do we start to unpack the approaches to generalize or approach new problems and solutions? Like many tools, artificial intelligence is an exciting tool, and it's still in speculation phase for what we will be able to get from it in advancing psychological knowledge. But, some effort testing the opportunities and limits is time well spent.

Which projects are absorbing you at the moment and what are your plans for the future?

The main research project that is dominating our attention is called the SCORE project. It's a project funded by DARPA. We're one of multiple teams involved in three technical areas: TA1, which is us, TA2 has two teams, and TA3 has three. The goal of the project is related to artificial intelligence, to see if we can create automatic indicators of the credibility of research claims. When you open a paper, each of the claims in the paper could have a score next to it: This one had 72, this one has 15, that one has 99. The machines would give an initial calibration of confidence that we can have in the credibility of the claim. This is a high aspiration, but pieces of evidence suggest that there is information we could extract from papers, and from the research at large, to help us assess the credibility in different findings as an initial heuristic. How much other evidence are these findings supporting? How does it fit with other claims? What fits that particular claim or that particular study? It's an exciting problem to try to solve, and the actual work that we're doing in our team is extracting claims from the literature.

We took 60 journals and extracted a sample of 50 articles from each year from 2009 to 2018 to create a database of 30,000 articles from the 60 journals. Then we take 10% of those as a stratified random sample, and we extract a claim from each article. We trace the claim from the abstract to a statistical inference in the paper supporting that claim. We then created a database of 3000 claims. TA2 teams evaluate the credibility of those claims with expert judgment and prediction markets giving each claim a score. TA3 teams are applying machines to try to assess the credibility of those claims. The machine gives them scores based on whatever information they can gather.

While all that is happening, we organize a thousand people in a massive collaboration to do replications of the substance of those claims as the ground truth. We shall see if the people in TA2 and the machines in TA3 can predict successful replication or not. This project will generate very useful data to study many questions in metascience and replicability. It's an extremely generative project, while simultaneously having a clear structure. And, the problem we are studying is an exciting problem.

The other problem we're really interested in studying is whether the various interventions we have introduced to improve the research process are working or not. We are running studies, for example, on Registered Reports to assess whether it's actually meeting the promise that we theorize, and whether there are costs, like the problem of creativity or conservatism that might emerge. Those would be very useful data to really help refine and improve the reforms—let's keep doing the things that are working and let's change or stop the things that aren't. This is the main areas of focus for me in the next few years.

It sounds promising. I know that most scientists avoid answering questions about the future. Nonetheless, I would like you to try and tell me how you imagine our discipline in 20–30 years' time?

Those are impossible questions, but I can at least answer what I hope for the discipline. My hope is that openness will be the default. And what I mean by open is the process of doing research, how I arrived at my claims. Maybe my approach included preregistration, or was pure exploration, or was something else. The goal is that you will be able to see my process from where I started to how I got to my claims. You will have access to the materials, the data and the things supporting that claim, to the extent possible. And, it will be easy for you to discover the evidence associated with that claim. Discovery tools will make it easy to say "here is my study, here are the 15 other studies that were done on that same problem."

If we have that, I think a lot of the other things that we'd like to see happen for the field will become more possible. It would be easier to solve the issue of publication bias, as we'll have meta-analyses that are much more credible for the state of understanding in this particular domain. More openness will make it easier to repeat things, we'll have a more routine ethic of replication, which will assess the credibility of particular claims and evidence. Having this will make it more likely that we'll start to tackle to the really hard problems, and actually make progress on them. If, in 20–30 years, we can get to a place where we can actively see how theories are developing in a systematic way in psychology, that would be great.

I have asked all my interviewees for advice they could give to young psychologists who are just starting their careers. You are my youngest interlocutor, but at the same time you have acquired experience that is completely different from that of others and you look at our field from a very broad perspective. What advice would you give to such young people?

To me, the most important things are to identify problems that one cares about, because it does matter to care. Just studying something because it's the thing I am studying, that won't keep people up at night in ways that I think are productive for doing scholarly research. Having questions that one is passionate about is a big value. The second thing is really thinking about what it means to me as a researcher to do the best possible research I can, and to prioritize those behaviors that are the ways that my science can be done best. I think it's very easy to fall into doing something because other people do it this way, other people say this is the priority. Of course, one needs to be attentive and responsive to the realities of the external rewards. But the satisfaction with the work that we do, I think, is largely internally driven. I won't be happy doing my research if I can't follow the values of how I think research should be done. That's important to prioritize, because if I can't live according to my values than I will just be constantly frustrated and feel like I am undermining my own goals.

Who would you point to as an example to follow?

Tony Greenwald. Tony has a long-standing and well-earned reputation of being hard to deal with, because he has a very clear idea of how he thinks science should be done. This makes him obstinate and impossible at times. I've published more papers with him than with anyone else. Each one of them has been torturous in its own way, but also extremely gratifying. The one thing that he does not sacrifice is wanting to be able to be confident in the claims he makes in a paper. So, he is constantly skeptical of his work, attacking it in different ways, so that we can believe in the output. I've tried through my career to emulate the things in his methodology, of doing the way it should be done. Tony is an emblematic example of that. And maybe I try to be a little bit nicer than Tony was about it. He isn't perfect, but he is fantastic, and I am so grateful for having been trained by him.

While working on this book I asked my readers to submit one question they would like to ask an eminent living psychologist. I received 30 of them. Would you agree to draw and answer one?

Sure, give me number 15.

What psychological idea is ready for retirement?

I'd say the idea that psychology is so different from other scientific disciplines that we can't use practices in other disciplines to inform how we can do ours better. I think this is partly a conceit, and partly low collective self-esteem that we think this way: "The reason that people see us differently, they don't respect us, is because we are totally different." But we aren't totally different. We are studying hard problems, perhaps the hardest among the sciences. But to then become so defensive that we can't look to other sciences to identify ideas and new practices to do ours better is a defensive stance that I think needs to die.

Selected Readings

Graham, J., Nosek, B. A., Haidt, J., Iyer, R., Koleva, S., & Ditto, P. H. (2011). Mapping the moral domain. *Journal of Personality and Social Psychology, 101*(2), 366–385.

Hawkins, C. B., & Nosek, B. A. (2012). Motivated independence? Implicit party identity predicts political judgments among self-proclaimed Independents. *Personality and Social Psychology Bulletin, 38*(11), 1437–1452.

Munafò, M. R., Nosek, B. A., Bishop, D. V. M., Button, K. S., Chambers, C. D., Percie du Sert, N., et al. (2017). A manifesto for reproducible science. *Nature Human Behavior, 1,* 0021. https://doi.org/10.1038/s41562-016-0021.

Nosek, B. A., Alter, G., Banks, G. C., Borsboom, D., Bowman, S. D., Breckler, S. J., et al. (2015). Scientific standards: Promoting an open research culture. *Science, 348*(6242), 1422–1425.

Nosek, B. A., Ebersole, C. R., DeHaven, A., & Mellor, D. M. (2018). The preregistration revolution. *Proceedings for the National Academy of Sciences, 115,* 2600–2606. https://doi.org/10.1073/pnas.1708274114.

Nosek, B.A., Spies, J. R., & Motyl, M. (2012). Scientific utopia: II. Restructuring incentives and practices to promote truth over publishability. *Perspectives on Psychological Science, 7*(6), 615–631.

Open Science Collaboration. (2015). Estimating the reproducibility of psychological science. *Science, 349*(6251), aac4716. https://doi.org/10.1126/science.aac4716.

Uhlmann, E. L., Ebersole, C., Chartier, C., Errington, T., Kidwell, M., Lai, C. K., et al. (2019). Scientific utopia III: Crowdsourcing science. *Perspectives on Psychological Science, 14,* 711–733.

References

365 days: Nature's 10. Ten people who mattered this year. (2015, December 24). *Nature, 528,* 459–467.

Baumeister, R. F., Vohs, K. D., & Funder, D. C. (2007). Psychology as the science of self-reports and finger movements: Whatever happened to actual behavior? *Perspectives on Psychological Science*, *2*(4), 396–403.

Cohen, J. (1962). The statistical power of abnormal-social psychological research. *Journal of Abnormal and Social Psychology, 65*(3), 145–153.

Doliński, D. (2018). Is psychology still a science of behaviour? *Social Psychological Bulletin, 13*(2). Retrieved from https://doi.org/10.5964/spb.v13i2.25025.

Hunter, J. (2000). The desperate need for replication. *Journal of Consumer Research, 28,* 149–158.

Simmons, J. P., Nelson, L. D., & Simonsohn, U. (2011). False-positive psychology: Undisclosed flexibility in data collection and analysis allows presenting anything as significant. *Journal of Social Archaeology, 22*(11), 60–80.

Wolins, L. (1962). Responsibility for raw data. *American Psychologist, 17*(9), 657–658.

14

Vikram H. Patel: Global Mental Health

I would love to live in a world where we can value our mental health in the same way we value our physical health.

Vikram H. Patel

T. Witkowski, *Shaping Psychology*,
https://doi.org/10.1007/978-3-030-50003-0_14

When today I recall the biblical story of David and Goliath, it appears to be a somewhat banal tale. For me, it lost its expressive power when I came to know Vikram Patel's life history, his achievements, aims and finally the man himself. Born in Mumbai, India, he was a sickly child, but he took up the challenge of mental illness, which was reaping its harvest across the world. He decided to create a situation whereby the poorest and most deprived in every corner of the world could count on some form of help. What is most surprising is that he has realized his aims with remarkable consistency.

As his mother Bharati said, Vikram Patel always wanted to be a doctor. But seeing his physical condition, people would laugh at his dream. How could a seriously ill boy (he suffered chronic asthma and allergy) who might not even live long enough to graduate medical school become a doctor? But Vikram was nothing if not determined. When he was admitted for the MBBS, he went to do his internship in Goa, where his friend Gauri, today his wife, gave him a form to fill out to apply for the Rhodes Scholarship. He sent it and was selected for an interview along with 15 other people. The panel was comprised of several leading personalities and the selection process went on for four days. In the end, he was selected as one of three Rhodes Scholars from India in 1986 (Bhatt 2015).

After completing his MBBS at the University of Mumbai, he went on to study at the University of Oxford, then finished his training as a psychiatrist in London. Next, he moved with his wife to Harare, the capital of Zimbabwe, to begin a two-year research fellowship at the national university. His objective was to find evidence for the view, then widespread among anthropologically minded psychiatrists, that what appeared to be depression in poor countries was actually a response to deprivation and injustice—conditions stemming from colonization (Patel 1997). The remedy in such cases, he believed, was not clinical treatments, but social justice. Instead of evidence for this, however, he discovered something quite different. It turned out that mental illness, and depression in particular, affects people in developed and developing countries in very similar ways. Not only that, as the first Global Burden of Disease reported at that time, they are the single largest cause of disability worldwide. In the poorest countries as well as the richest, and at every socioeconomic level in between, mental disorders were the greatest thief of productive life. As Patel said in one interview, "Zimbabwe was an eye-opener. It set me on a course I have been on since then" (Jain 2015).

This course is to find out ways to help the mentally ill worldwide, regardless of where they live or their material circumstances. He accomplishes this goal through advocacy, research, and teaching. He does so with such determination that his efforts have been recognized by his inclusion in the TIME 100 list of the most influential people in the world in 2015 (Van Dahlen 2015). He has been awarded the Chalmers Medal from the Royal Society of Tropical Medicine & Hygiene (UK), the Sarnat International Prize (the National Academy of Medicine, USA), an OBE from the UK government, the Pardes Humanitarian Prize (the Brain & Behaviour Research Foundation, USA) and the Canada Gairdner Global Health Prize. He was elected a Fellow of the UK Academy of Medical Sciences, awarded an Honorary Doctorate from Georgetown University and awarded a Wellcome Trust Principal Research Fellowship.

Although Vikram Patel is a qualified psychiatrist, he rejects complicated psychiatric diagnostic categories which are unsuited to community contexts. He was a professor at the London School of Hygiene and Tropical Medicine, yet he somewhat feels uncomfortable in the academic world isolated in its ivory tower.

A brief profile does not give enough space to list all his achievements. However, some of them cannot be omitted. These certainly include writing and publishing in 2003 the book *Where There Is No Psychiatrist*, a mental health care manual primarily used in developing countries by non-specialist health workers and volunteers (Patel 2003).

On his request, the publisher entrusted the translation rights of his book on one condition—that they could use it as long as it was distributed free of charge. This book has since been translated into fifteen languages and helps spread the simple yet profound idea of mental health in over 70 countries of the world for all. A second, much-revised, edition was published in 2018 and, this time, the digital version of the book can be downloaded at no charge.

His work is firmly grounded in science, using epidemiology, social science research methods and randomized controlled trials. He has used them to create and test community treatment protocols for everything from depression to chronic schizophrenia in India. The product of this work includes over 300 scientific articles published in peer-reviewed journals.

Vikram Patel is the co-founder and former director of the Centre for Global Mental Health, London School of Hygiene and Tropical Medicine (LSHTM). He was also the co-founder of the Centre for Control of Chronic Conditions at the Public Health Foundation of India, and co-founder of Sangath, an Indian NGO (where he continues to serve as a member of the Managing Committee) which has pioneered task-sharing experiments in the areas of child development, adolescent health, and mental health. Sangath won the MacArthur Foundation's International Prize for Creative and Effective Institutions in 2008 and the WHO Public Health Champion of India award in 2016 and is now ranked amongst India's leading public health research institutions. After serving as Professor and Wellcome Trust fellow at the London School of Hygiene & Tropical Medicine for a decade, he became the first Pershing Square Professor of Global Health at Harvard Medical School in 2017 where he launched the GlobalMentalHealth@Harvard initiative. He now splits his time between Boston and India.

He has served on three WHO Committees (Mental Health; Maternal, Child and Adolescent Health; Independent High Level Commission on Non-Communicable Diseases) and on four Government of India committees: the Mental Health Policy Group (which drafted India's first national mental health policy), the National Rural Health Mission ASHA Mentoring Group, the National Human Rights Commission Core Committee on Health and the Technical Advisory Group of the Rashtriya Kishor Swasthya Karyakram (India's national adolescent health program).

Professor Patel, as you are engaged in the area of mental health, the subject of both research and practice by a huge number of psychologists, I would like to start by asking you a few general questions about the subject. On many occasions, you have talked about how, from the perspective of primary healthcare workers, a lot of psychiatric categories do not make sense because they fail to reflect the reality of that field. Can we essentially toss our manuals like DSM or ICD into the shredder?

We need to design classifications which are meaningful to the people who use them and those who are classified as a result of them. In that sense, DSM and ICD are neither useful to the front line providers in primary health care, nor are they meaningful to most people who receive a diagnosis. We need to think of psychiatric classification as being pragmatic. At the level of the community and primary care, what really works are syndromes of distress, rather than the hundreds of different diagnostic categories that we have invented, which may be more useful for those working in a mental health specialist setting. The reason is that most psychiatric conditions are really dimensions, every individual with a psychiatric condition lying somewhere on a spectrum, and that when we classify people essentially, we are artificially or arbitrarily dividing the spectrum into binary categories. That is certainly useful for certain purposes. For example, it might be very useful for insurance purposes or for applying for welfare benefits. They can also sometimes be useful for treatment decisions, but often they are not very useful for people who are delivering psychological and social interventions or for those who receive the diagnoses. For example, such binary categories imply that those who are classified with a 'disorder' are different from other people who might also be experiencing very similar sort of experiences, but do not meet the arbitrary criteria of the diagnosis.

Is this also an effect of the medicalization of our life?

Yes, this is partly true, but it is important to know the history behind this. Up until the 1960s, it was very uncommon to have so many different diagnoses. We used to have very broad syndromes, like neurosis and psychosis, but in the 1970s there was a push in psychiatry to become more like its brethren in other medical specialties. These specialties were heavily influenced by diagnostic systems, which in turn were influenced by infectious diseases, where the diagnosis followed a causal agent. For example, you have tuberculosis that replaced a diagnosis of a cough—you could have many different infections causing a cough, but the diagnosis was not a cough, but TB or influenza, etc. However, because psychiatrists did not have any such causes, what they decided to do was to create diagnoses based on what they were observing in their clinics and hoped that this would, one day, through research using those diagnostic categories, help us to uncover a cause that was specific to those diagnostic categories. However, 40 years later we now know that that is not the case. To be fair, historically there was a good reason for applying these diagnostic categories. There was a genuine belief that the causes could be discovered by actually classifying people according to their clinically observed syndromes, because it was assumed that these syndromes would be very distinct from one another in terms of pathology. Today we know that's not the case, and that is why we need to rethink how diagnoses are applied to the field of mental health. It's very clear that the model that works so well for infectious diseases is not going to hold true for mental health problems.

Another general question. The World Health Organization estimates that more than 350 million people around the world are currently suffering from depression, which accounts for almost 5% of the population. Worldwide more than a third of people at some point in their lives meet all the diagnostic criteria for at least one mental disorder. These data suggest that there is a mental illness pandemic of unprecedented magnitude in the history of mankind. What is your explanation for these figures?

I cannot agree that there is a pandemic, and cannot say that things are getting worse, if only because we have no real knowledge about the prevalence of mental illness before the diagnostic system came into play, which is only 30 years ago. That said, I don't think that it would be an overstatement to state that the burden of mental illness is very high. If you think of the full range of mental health problems—from problems to do with social communication in childhood, learning and behavior in adolescence, self-harm, use of harmful substances, feeling extremely miserable for extended periods of time, trauma through violent events, having sustained psychotic experiences—and you put them all together, it is hard to find anybody who actually has never experienced mental health problems at some time in their life. But to me this is part and parcel of the human experience. Our mental health is like our physical health. If I were to ask the question, how many people have ever had a physical health problem in their lives, would you be surprised if you heard the number 100%? Of course not, because everybody has a physical health problem, especially if you consider that physical health problems extend from, for example, a common cold all the way to lung cancer. In the same way, if you think of mental health problems as a spectrum, ranging from acute distress, for example because your wife has died after 30 years of marriage, all the way through to actually being depressed in a sustained way for months and months after she died, then almost everybody has had a mental health problem in their lives. Of course, this does not mean that everyone would have had a 'diagnosis' of a mental disorder! I think we should move away from numbers because they are misleading, and they can actually make us focus on the wrong issue, which is whether it's 349, 350, or 351 million. Who cares about such numbers when these diagnoses are not really grounded in science? Where we really should be asking the question is, what proportion of humanity experiences difficulties with their mental health at some point in their lives? And if we asked that question, I think the answer is 100%.

What about the practice of over-diagnosis?

I agree, that's a huge problem for me, and that is consequent on the fact that we think that every time our mental health is effected, it means we have a medical problem, and that requires us to see an M.D. or Ph.D. with specialist training and experience, and pay them significant amounts of money to get a diagnosis and, most often, a drug prescription. I completely reject that approach. What the dimensional approach to mental health tells us is that, as with physical health, a huge amount of that spectrum can be managed with appropriate changes that one individual can make to their own lives, through adequate information, hope and informal approaches of support and care and, as we have shown in our work, through community health workers who are able to deliver brief psychological and social interventions. For example, let me give you an analogy with heart disease. The majority of people with heart problems can actually manage their health through lifestyle change, and if they have a cardiovascular event, the majority of lives can be saved by a first responder being trained in CPR, so only a small proportion of people actually need to see a cardiologist or go through a cardiac procedure. We need to apply similar dimensional approaches to thinking about the prevention of mental health problems, the management of acute mental health distress, the front-line response to mental health problems and, finally, the specialist response to refractory and persistent mental health problems.

To some extent, your activities undermine traditional beliefs about the effectiveness of professional treatment, as through your research and practice you have proven that with the help of simple tips read in a book or given to them during a short online course, people without any qualifications can contribute to a significant improvement in the mental health of people suffering from serious disorders. Such results are also consistent with what Raj Persaud (1998) **wrote in his book *Staying Sane*, showing that unskilled or novice healthcare professionals are often more effective than professionals. How do proponents of traditional, highly professional psychiatry and psychotherapy react to this?**

I think that many of them react in the same way that physicians would react when they hear that a physician assistant or a nurse is able to manage the majority of physical health problems. However, it is extremely important not to see this as a debate about who is more effective. Instead, it is more useful to ask the questions—what are the skills are needed for particular kinds of problems, who is in the best position to offer these skills, and what does it take to achieve that degree of skill? For example, one would never ever question that the skills that are needed to manage an acute psychotic episode are quite different to the skills that are needed to help somebody who has lost a loved one and is suffering from acute bereavement to recover. To try and suggest that all of these are the same skills is a mistake, just as it is wrong to assume that all of these individuals need a biomedical approach. I think you need both psychosocial and biomedical approaches in the right proportion, tailored to the needs of the person.

As shown in the report you and your colleagues (2010) **published in *The Lancet*, the efficacy of some of the forms of help you propose is astonishing. Doesn't this directly explain the demand to simplify or at least review existing psychotherapeutic methods, whose effectiveness lags far behind what is achieved by non-professionals in developing countries?**

Absolutely. I think that the work my colleagues and I have done in global health over the last 15 years clearly demonstrates two things. Firstly, that non-specialist providers such as peer support workers, community health workers and lay counselors can be equipped with the necessary skills to help people with mental health problems recover, and secondly, that the contents of these interventions are relatively simple, brief and technically feasible to deliver in routine care settings. Secondly, what this evidence also shows is that there is a universally applicable scientific foundation to explain the nature of mental health problems, and therefore to also explain how specific psychosocial techniques that are addressing those mechanisms can help people recover. To me this is also very exciting. It shows that psychological reaction to adversity is a universal human response, that this follows very similar pathways, and therefore the kinds of techniques that we have developed in order to address them, no matter which tradition they emerge from, can be universally applicable. Diverse traditions of psychology, sociology and even spiritualism are often acting through similar pathways to help people recover. I should emphasize here that I am focusing mainly on mood and anxiety problems, which are a distinct group of conditions from psychoses. I think psychoses, on the other hand, are much more heavily determined by genetic and biological factors, and they are also much rarer conditions. There is no question in my mind that people with psychoses do benefit from medication as well as psychosocial approaches, but mood and anxiety problems, which are the biggest category of mental health problems, are to a large extent influenced by environmental factors. In that group of individuals, current evidence suggests that psychological and social interventions are far more likely to produce lasting benefit than pharmacological approaches.

In your public statements, you often stress that when you started your career as a psychiatrist, you were convinced that depression is a disease of Westerners. Today you admit that you were wrong, and that people in all cultures suffer from depression. However, I am curious if you perceive certain mental disorders that are present in one culture and not present or very rare in other cultures. Such cases are described by James Davies (2013) in his book *Cracked: Why Psychiatry Is Doing More Harm Than Good*, showing that some disorders which are alien in a given culture, such as eating disorders, appear in the East as a result of the cultural pressure exerted by Western civilization. He called this influence "psychiatric imperialism." What is your view on psychiatric imperialism?

I think these are two completely separate issues. The influence of culture on our mental health is a global issue. I completely agree that culturally determined attitudes toward body shape are critically important environmental influences on dieting behaviors, and the emphasis on thinness is a major reason why eating disorders were so much commoner in European society. That said, as the world globalizes those attitudes are now finding their way into dieting behaviors in non-Western societies. Similarly, you also see other kinds of cultural factors operating in non-Western societies, which lead to mental health problems—for example, in India, there is a psychosomatic syndrome affecting young men characterized by severe anxiety about semen loss related to cultural attitudes about virility. I think the influence of culture on our mental health is a well-established and universal phenomenon. Psychiatric imperialism, on the other hand, is about the application of a biomedical psychiatry on cultures in which there is no history of psychiatry. Thus, it is considered that psychiatry is a product of Western culture, and it is imperialistic to apply it to non-Western cultures. I don't agree with this at all. I think psychiatry is a discipline of medicine, and in the same kind of way that the other disciplines of medicine have been applied successfully around the world I believe that psychiatry can also be applied around the world. However, where I do have a problem with psychiatry is its failure to recognize that diagnostic categories do not travel across cultures, not only around the world, but even in European cultures, in particular the narrow pharmacological perspective of addressing mental health problems. There is a need for the reformation of psychiatry so that it is grounded much more in both neuroscience, in understanding how mental health is produced in the brain, and in epidemiological sciences, to understand how our social environment influences our mental health, and to apply these in a person-centered way. That is to say, there isn't one size that fits all, and rather than single-mindedly follow a protocol that allocates a person into a diagnostic category which then leads to a treatment plan, we need to understand the context in which that person's mental health problems are occurring, and provide interventions that are tailored to that person. To me, this is a universal need rather than occurring only in one culture or another. The application of the model of diagnosis that we talked about earlier is without thought to the person or the context, and I think that is a problem everywhere.

In his book *The Mind Game: Witchdoctors and Psychiatrists*, Erving Fuller Torrey (1972) was probably the first to attempt to convince the Western cultural community that the actions of witch-doctors and local healers in other cultures are the full-fledged equivalent of our psychiatry and psychotherapy. Your activity is done in places around the world where similar services are quite well-developed and available, and it could constitute a substitute or even an equivalent of professional psychiatric care. Do you see any opportunities for working with these people? Do witchdoctors come to people's aid?

I wouldn't call them witch doctors. I call them traditional medical practitioners or traditional healers. I certainly see that in some cultural contexts they play a very important role, particularly for chronic conditions, which included not only mental illness but also a range of other conditions, from allergic disorders to diabetes, etc. I do believe that biomedicine is not the only or dominant form of health care that is required, especially as most of these chronic conditions are a product of interaction with an environment. I see the role of other therapeutic traditions as being very relevant, but just as with biomedicine, there is also a risk of abuse in these other traditions of practice. Most of these traditional systems operate without any scrutiny, monitoring, or quality control, and almost entirely in the private sector. Some of their practices can be very abusive, such as chaining, whipping and branding, and many blame the person of their family for the sickness. I think that one should not romanticize and valorize these alternative systems simply because they have come from another culture. There is a danger of the "happy native" concept that has pervaded anthropology for a long time which postulates that the so-called natives are happy with the living in their jungle and wearing no clothes and the so-called Western way of life is always a destructive or malevolent force. I think that is also a fundamentally racist concept, and was at the heart of the rationale which European nations used to systematically abuse its colonized peoples. I think we have to be very careful about that kind of thinking and the history behind it. It is a very fine line between that and the idea that all traditional medicine is beautiful and wonderful, because I have seen traditional medicine as being anything but beautiful and wonderful.

In the places your help reaches, mental illness is often confused with possession or spells. Doesn't that make it more difficult to diagnose people with mental illness, the difference between their expectations of how to deal with them and what they expect in reality?

I don't think there is any one mental illness. It's rare for people with mood and anxiety problems to think they're being possessed, and most people really see their experiences as a result of social suffering, which of course is exactly what epidemiology also teaches us. Possession states are mostly associated with extreme anxiety or psychosis, and in those situations, it is important to remember that the possession belief is very much rooted in the fact that the spirit and the mind have always been seen as being interconnected in most societies of the world, including Europe. It was only after the Enlightenment, which separated religion from medicine, and the emergence of biomedicine about 150 years ago that spiritual matters become the realm of the church, and medicine became the realm of the biomedical practitioner. But in most parts of the world, of course, that association of spirit and the mind continues to this day. Indeed, you can also see this in the west—for example, if you go to any of the evangelical churches in Europe or North America you will see a lot of intermingling of spirit and mind, and witness many people coming for spiritual healing who might also be considered as suffering from a mental health problem. I think this is a very rich expression of how people understand their mental health problems and the kind of help they seek, and as long as the practice is not going to be harmful to the person, either economically or because of the loss of dignity or autonomy, I think all of these alternative forms of healing are acceptable as the person may ultimately benefit from.

I got to know the problem of possession quite well in my country in the middle of Europe. Poland is one of the leaders in exorcisms in the world. But let's return to the developing world. Leo Igwe, the Nigerian human right advocate and humanist, told me about how people accused of witchcraft, including children, are persecuted in Africa. What does the stigmatization of mentally ill people look like,

when its roots are perceived are being in possession or spell? Is it different than in the West?

I have read about women being labeled as witches in India as well, but most often these labels of are not to do with mental illness but rather with misfortunes that are happening in the community, and then blaming a particular individual for being a witch. Here, the link with mental illness is much weaker, it's much more to do with misfortune and other social factors, such as those related to gender. For example, in India someone is often accused of witchcraft if in a particular family three children have died one after the other, leading to the conviction that there is some kind of bad spell or evil spirit in the house. I think the stigma associated with mental illness has to a large extent been the result of the way people with mental illness have been treated, and that could be either in traditional healer shrines, where they might be chained or whipped, or over the last hundred years in most countries through asylum psychiatry. If you think about the way people with serious mental illness are treated, that they are taken away by the police, incarcerated, and no longer treated with dignity, but as some kind of sub-human creature. They are put into hospitals where they are naked, raped, and given ECT as punishment for "bad behaviour." If this is what passes for mental health care, then it is no surprise at all that people don't want to seek mental health care and be stigmatized with a psychiatric label. People mock and joke about mental illness, because over hundreds of years we have done exactly that as part of our societal and professional response to people with mental illness.

The working methods you promote are designed for areas of the globe inhabited by people deprived of access to professional mental health services. Nevertheless, in developed countries a large part of the population (estimated to be as much as half) also does not have access to such services. In your opinion, could the methods you have developed also be applied in Europe and North America?

Yes, absolutely, and in fact I am already involved in such efforts. With colleagues in Canada and the USA, we are now applying the same methods of brief, technique focused, psychological therapies to be delivered by front line workers such as community health workers and nurses. There is a huge potential to reform the architecture of mental health care, so that for the majority of people with mental health problems their first level of care is not in a hospital, a clinic or a mental health professional setting but either in their own homes or in primary health care, with treatments that are brief and seek to empower that person to be able to learn the skills and techniques and to give themselves time and space to recover, rather than medicalize them, apply meaningless diagnostic categories and reduce them to patients passively taking a prescription.

For some time now I have been interested in the possibilities offered by computer-assisted therapy and the use of AI in psychotherapy. Achievements to date in this area seem promising—inexpensive help readily available in even the most remote places in the world, as long as there is an Internet connection. Does your program of help for the most needy and poorest people in the world account for these possibilities? What prospects do you see for them?

I have a very different view of digital technologies. I think most people who are distressed don't want to sit with a computer to get better. Remember that 15 years ago, there was much excitement around the digital revolution offering the potential of apps based on psychotherapy which people could use themselves so that we could actually remove the need for a therapist completely. We now know, 10 or 15 years later, that none of the thousands of apps which were developed have actually been successful or commercially viable. Companies will say they had a million people accessing the app, but then they will never tell you how many people actually completed more than one session because those numbers are very dismal. What we do know now is that the vast majority of people do not want to sit with an app alone, and now the field is moving toward blended approaches, recognizing the need for a skilled person who can use an app, where appropriate, to support recovery. The person doesn't have to be an M.D. or Ph.D., but somebody who can assist you with their knowledge to help you recover. The digital world is then something that you use as a support for that, rather than a replacement of the skilled provider. One example of the way I use digital applications in my work is to build a platform for training providers to learn psychological treatments, so we can address the barrier of the old-fashioned face-to-face training and supervision, which is one of the major reasons why none of these brief evidence-based treatments have gone to scale. Another example is using digital platforms for supporting the quality of the care that is delivered and for providing supervision. Such provider and health system facing apps, combined with patients facing apps, is perhaps something I would feel very excited about, but just simply giving a patient an app with AI makes me very pessimistic. The latter may help a handful of people who are very motivated and fairly adept at using such digital offerings, but for the majority of people with mental health problems, I see them using these apps only as a supplement to skilled providers.

Although you are a scholar educated at Western universities, you enjoy the capacity to look at mental health, psychiatry and psychology from a perspective quite different from that of most of the readers of this book, and it would be a waste not to use it to paint a more detailed picture of our field. Looking at psychology through the eyes of someone who tries to use its findings to help mentally ill people in the poorest regions of the world, which of its achievements do you consider to be particularly significant and beneficial?

I think psychology is probably the most important science of all, when it comes to understanding mental health and mental illness. It is certainly a lot more scientifically grounded than psychiatry is. In fact, most of the ideas that we are talking about right now have a rich and long history in psychological science. For example, the dimensional model of understanding mental health, such as concerning with cognition and different emotional states, has a very rich history in psychology, and it is psychiatry that took these ideas and created biomedical categories out of them. I think of social and psychological sciences as the fundamental foundations of mental health science, alongside neuroscience, and it is clinical psychiatry that in many ways is out of step. I would suggest that practitioners of psychology should follow the original thinking that lay at the heart of psychological science, rather than trying to ape the psychiatric practice of mental health. Among its greatest contributions are the understanding of cognitive and behavioral foundations of mental health, the interpersonal dynamics through which mental health is produced and the understanding of how our social environment affects the brain, particularly in the early years of life. Such discoveries, and many other aspects of psychology, form the essential science of mental health, alongside functional neuroscience. For me, actually one of the real disappointments these days is the desire expressed by some psychologists to be psychiatrists. While I do appreciate that to a large extent this desire is driven by commercial interest, but I still find it amazing, and even amusing, that there are some psychologists who want prescribing powers despite the fact that the most powerful mental health interventions that we have today, the ones that are most acceptable to consumers, the ones that have the least side effects and that have the most sustainable long-term effects, are actually those that are derived from principles of psychological and social interventions.

Perhaps it's also a pressure from patients. They expect a psychologist will prescribe them something.

I actually don't think so at all, because if you look at the literature about people's choices about what kind of care they receive when they're offered a choice, there is an overwhelming preference for psychological and social interventions, especially when they are dealing with mood and anxiety disorders.

What therapeutic methods have proven themselves in the difficult conditions you work in?

I think that the most effective ones in the psychological realm are to do with behavioral activation, which is relatively simple technique to teach both the provider and the patient. Examples include behavioral activation that is primarily used for depression and exposure which is used for anxiety. Motivation enhancement is a very effective technique for promoting healthy behaviors. Problem-solving is a very effective technique for helping young people, especially for executive decision-making and emotional regulation. Another set of approaches which come outside the narrow domain of clinical psychology are focused on social work practices, for example enabling people to nurture social relationships through befriending interventions or supporting people with severe mental disorders to be included in mainstream spaces, such as in school, the workplace, the family environment or the neighborhood.

Which methods have proven themselves to be utterly useless or even dangerous?

The one approach that I think is the most dangerous and useless is the incarceration of people with mental illness. It damages everyone concerned, but most of all the person with mental illness, because it robs them of their freedom and their autonomy. Typically, when they are incarcerated against their will, whether in hospital, at home or in a prison, they are robbed of most or all of their fundamental human rights. But this practice also robs the providers and the mental health professions of their dignity and respect. And so I think the one practice that I would like to see forever ended is the incarceration and involuntary treatment of people with mental health problems. Then there are a whole range of psychiatric practices that are not evidence based, such as long term psychoanalytical therapy or regression therapy, which have no scientific basis. Typically, such practices are offered only by private practitioners, who have completed a course somewhere and charge huge amounts of money. To me, these are people who are preying on vulnerable people with mental health problems, and I think we should reject this kind of charlatan practice. Part of the problem is that mental health care systems are so weak everywhere that it offers a much larger space for charlatans with no scientific base to establish a foothold. You would never see this, for example, in cancer therapeutics, and that really is a reflection of how weak evidence based mental health care is in most parts of the world.

One of the very important issues in contemporary clinical psychology is the gap between science and psychotherapy. By that, I don't mean pseudo-therapy, which has nothing to do with science, but those modalities that are commonly considered evidence-based. In one study (Jonsson et al. 2014), it was found that only 3% of all studies on the effectiveness of psychotherapy included monitoring of negative side effects. In another, it turned out that most of the research is conducted by researchers who do not declare a conflict of interest (Lieb et al. 2016). Yet another meta-analysis showed that only 7% of all studies contain convincing evidence confirming the effectiveness of psychotherapy (Dragioti et al. 2017). The picture that emerges from these and many other works is not very optimistic. What are your views on this subject?

I share the concerns that a lot of the science in both psychotherapy and pharmacotherapy has been produced by people with a vested interest, which they often don't disclose, and often these findings are not replicated, and this is a real problem in our field. It goes back to what I've mentioned earlier about the fact that there is weak evidence based science in most mental health practice today, and that allows fertile ground for people to step in and offer all kinds of therapies, because they are likely to say there is no evidence base to even back up what the formal system is providing. This is where, I think, what global mental health is doing is really exciting, because it is not using branded therapies. Instead, it is generating evidence on the utility and effectiveness of simple techniques that can be used by ordinary people to help themselves or other ordinary people who are going through mental health problems. It is deprofessionalizing psychological science and making it available to people in a democratic way. You also probably know that most people who claim they are psychotherapists don't actually practice evidence-based psychological therapy, but practice an eclectic mix of whatever they think is needed. This is probably the reason why community health workers probably do better than psychotherapists, because they stick to the simple approaches they have been taught to deliver.

Today everybody is talking loudly about the crisis in psychology. And I don't mean only the replication crisis. It's a much wider issue. In your opinion, which areas of psychology have we neglected? What should we focus on so that our field does not merely remain an exciting hobby allowing us to describe our behaviors and motivations in ever finer detail, but rather helps to solve the most present problems of humanity?

I think what psychology should be doing is returning to its roots of attempting to understand and study the dimensions of mental health and personality. I consider, for example, the whole area of neuro-cognition and psychopathology as fundamental psychological science. These are the areas that we should be focusing on, and trying to understand how our social worlds are influencing our psychological state, and furthermore connecting this with the growing and exciting developments in circuit and network neuroscience. I think we should stop thinking about diagnostic categories and revert to the original, historic scientific foundations of psychological science. We should embrace exciting new fields, such as cognitive and network neuroscience. Digital tools will also play a role, but the digital tools of the future will not treat diagnoses such as depression or ADHD, but help people learn and master techniques to deal with impairments or difficulties in dimensions of psychological function. For example, in the case of ADHD, tools which can help children improve their attention span, and reduce their impulsiveness. That's where psychology should be heading.

Readers interested in psychology rarely read about global mental health problems. You have achieved a lot in this field. Which of these achievements are you most proud of and why?

It is really disappointing that few people in psychology talk about global mental health, because most of global mental health is about psychology. It is necessary for counseling and clinical psychology to recognize how influential they are to the way the world is beginning to understand and respond to mental health problems. I think the most exciting things emerging from global mental health are the deconstruction of diagnostic categories into dimensions of psychological function or experience, and also the application of deconstructed complex psychological therapies into specific elements or techniques that can be used to help people with different forms of mental health problems recover.

What problems are you trying to solve and what questions are you trying to find answers to these days?

Firstly, early child development. I am engaged in programs to try and characterize the development of very young children using dimensions of neuro-cognition, rather than trying to classify them into diagnostic categories. The second area is working with adolescents, and again, moving away from diagnostic approaches to study and address difficulties in emotional regulation which lead to impulsive acts and behaviors such as self-harm. The techniques that can be used are based on the principles of problem-solving, to help young people pause before they act impulsively and be able to identify and deal with stresses in their everyday lives. The third is to disseminate evidence based psychological treatments using digital platforms for training front line workers, to complete competency assessments and to master their delivery.

What other questions in global mental health that you would like to be able to answer in the next few years?

I think one of the other very important areas that I am passionate about, and which I am increasingly seeing myself drawn into, is how to end coercion in mental health care. How do we actually empower people with the most acute mental health problems to be supported in decision making in their own interest, rather than to substitute decision making to others. I would like to see all forms of coercion and violence in mental health care completely eliminated, and to me that is an aspiration and a moral mission. Another area which I am also very interested in is to begin to better understand how poverty and deprivation affect our mental health, so that we can interrupt the pathways between living in conditions of adversity and suffering the mental health consequences of these adversities. In that way we can build capabilities in people who are very disadvantaged to lead lives that are more fulfilling and rewarding, even if we can't necessarily change the upstream social and structure determinants of poor mental health.

What is your biggest dream related to what you do?

I would love to live in a world where we can all truly value our mental health in a way we value our physical health. To do that, we have to abandon the idea that mental health is about diseases, disorders and specialists, and recognize that mental health is something that accompanies each of us at every single point in our lives, as long as we are alive, whether we are awake or asleep. Also to value our mental health is to be conscious of it, to be able to talk about it freely and comfortably, and be empowered with the knowledge and tools to care for it. This is the world that I really envision and aspire to.

This book will mainly be read by psychologists. What message would you like to convey to them?

My message is, became a practitioner of a person's mental health, not their mental illness.

Is there anything I haven't asked you about during this conversation that you would like to mention?

I want to reiterate one more point and that is that all mental health problems, without exception, have a social dimension, and I think we must address the social world that the person occupies, as much as their internal mental experiences. I think that only then can we have a truly person-centered approach to mental health, with a long-lasting impact. The problem in the field of psychology is that it has focused entirely on the inner world, and assumed that the outer social world is the realm of another discipline, for example, social work. But I think that is a mistake.

It is a limitation of psychology not to also be able to at least understand and acknowledge this, and assist the person to re-order their social world as much as possible, because to ignore the social world is to ignore much of the picture.

Selected Readings

Patel, V. (1998). *Culture & common mental disorders in Sub-Saharan Africa.* Hove: Psychology Press.

Patel, V., Aronson, L., & Divan, G. A. (2013). *School counsellor casebook*. Delhi: Byword Publishers.

Patel, V., Cohen, A., Minas, H., & Prince, M. (Eds.). (2013). *Global mental health: Principles and practice*. Oxford: Oxford University Press.

Patel, V., & Hanlon, C. (2018). *Where there is no psychiatrist*. Cambridge: Cambridge University Press.

References

Bhatt, S., (2015, September 11). The lady behind one of *Time's 100* influential people. Retrieved from https://www.rediff.com/news/special/the-lady-behind-one-of-times-100-influential-people/20150911.htm.

Davies, J. (2013). *Cracked: Why psychiatry is doing more harm than good*. London: Icon Books Ltd.

Dragioti, E., Karathanos, V., Gerdle, B., & Evangelou, E. (2017). Does psychotherapy work? An umbrella review of meta-analyses of randomized controlled trials. *Acta Psychiatrica Scandinavica, 136*(3), 236–246.

Fuller, M. D. E. (1972). *The mind game: Witchdoctors and psychiatrists.* New York, NY: Emerson.

Jain, A. (2015, April 19). *The mind's a big place*. Retrieved from https://fountainink.in/qna/the-minds-a-big-place.

Jonsson, U., Alaie, I., Parling, T., & Arnberg, F. K. (2014). Reporting of harms in randomized controlled trials of psychological interventions for mental and behavioural disorders: A review of current practice. *Contemporary Clinical Trials, 38*, 1–8.

Lieb, K., Osten-Sacken, J. V. D., Stoffers-Winterling, J., Reiss, N., & Barthet, J. (2016). Conflicts of interest and spin in reviews of psychological therapies: A systematic review. *BMJ Open, 6*(4). Retrieved from https://bmjopen.bmj.com/content/6/4/e010606.

Patel, V. (1997). *Common mental disorders in Harare: A study in the "new" cross-cultural psychiatry*. Ph.D. thesis, University of London, london.ac.uk.

Patel, V. (2003). *Where there is no psychiatrist: A mental health care manual.* London: Gaskell.

Patel, V., et al. (2010). Effectiveness of an intervention led by lay health counsellors for depressive and anxiety disorders in primary care in Goa, India (MANAS): A cluster randomised controlled trial. *The Lancet, 376*(9758), 2086–2095.

Persaud, R. (1998). *Staying Sane: How to make your mind work for you.* New York, NY: Metro Books.

Van Dahlen, B. (2015, April 16). The world's 100 most influential people. Vikram Patel. *TIME.* Retrieved from http://time.com/3822953/vikram-patel-2015-time-100/.

15

Daniel Kahneman: Decision Making, Adversarial Collaboration and Hedonic Psychology

I firmly believe that reducing misery is a more important objective for society than enhancing happiness.

Daniel Kahneman Photo by Andreas Weigend. CC BY 2.0; https://creativecommons.org/licenses/by/2.0/; photo has been cropped

T. Witkowski, *Shaping Psychology*,
https://doi.org/10.1007/978-3-030-50003-0_15

If anyone would have just cause to complain about their childhood, it would be Daniel Kahneman. He was born in Tel Aviv, the son of Lithuanian Jews, but he spent his childhood in France, which was occupied by German fascists when he was barely six years old. The first blow was the internment of his father. Although after some time his father managed to escape from the internment camp, the persecution of the Jews meant that the Kahneman family were forced to flee to Juan-les-Pins on the Côte d'Azur. They didn't settle there for long. When the Allies took North Africa, the Germans withdrew their forces to the south of France, forcing the Kahnemans to flee once again, moving from village to village. Eventually they ended up living in a chicken shed adapted for human habitation. Kahneman says of those times that he felt like a hunted animal—"We had the mentality of rabbits" (Shariatmadari 2015). Young Daniel concluded that God, to whom he prayed, must have been exceptionally busy in those times and unable to deal with major requests, so he thought that the most effective form of prayer would be to ask God for just one more day. Therefore, every day he prayed for just one more day of life ….

However, Kahneman does not complain about those times, nor does he seek self-justification in his experiences. Quite the opposite—he denies that the experience was traumatic. Perhaps it is precisely because of his attitude of not looking back that he was able to embark on an intellectual journey, guided by his unbounded curiosity. This journey eventually led him in 2002 to the Swedish Academy of Science, who awarded him the Sveriges Riksbank Prize in Economic Sciences in Memory of Alfred Nobel. Today, he is one of the world's most influential psychologists.

Shortly after the liberation of France, Kahneman moved to Israel where he decided to study psychology. He got his bachelor of science degree with a major in psychology, and a minor in mathematics from the Hebrew University in Jerusalem. In 1958, after completing military service in the Israel Defence Forces, he went to the USA to study for his Ph.D. in psychology from the University of California, Berkeley. In the spring of 1961, he wrote his dissertation on a statistical and experimental analysis of the relations between adjectives in the semantic differential. After completing his Ph.D. studies Kahneman went back to Israel and began his academic career as a lecturer in psychology at the Hebrew University of Jerusalem in 1961. His early work focused on visual perception and attention.

However, Kahneman began the greatest adventure of his life at the end of the 1960s, when he began co-operating with Amos Tversky, a younger colleague from Jerusalem. He speaks of this meeting as a "magical experience." Together, Kahneman and Tversky formed that rare kind of team, where opposites complement each other perfectly, and they published a series of seminal articles in the general field of judgment and decision making, culminating in the publication of their prospect theory in 1979 (Kahneman and Tversky 1979). It is chiefly for these works that he was awarded the Nobel Memorial Prize in Economic Sciences. Kahneman and Tversky would continue to publish together until the end of Tversky's life, but the period when they published almost exclusively together ended in 1983.

The culmination of his work on decision-making processes is his bestseller *Thinking Fast and Slow*, published in 2011. In it, Kahneman explains how the brain works, by describing two processes which together decide our way of thinking. The first is fast, intuitive and emotional, whereas the second is slower and operates in a more considered and logical way. Although the author shows these two brain mechanisms to be lie uneasy together (because they are counterintuitive), the book has enjoyed enormous popularity among millions of readers and has received many awards.

One of Kahneman's rarely mentioned (not to say forgotten) achievements is in developing a procedure of so-called adversarial collaboration, which is an answer to the absurd (in his opinion) method of conducting scientific debate, where an author replies to his critics in kind, which, instead of leading to consensus, simply further polarizes the position of the adversaries. Adversarial collaboration depends on attempting to resolve differences of opinion through jointly conducted research. In this method the adversaries together develop a research procedure which they think will resolve the dispute. Kahneman has often expressed the hope that this method will become part of his legacy (Kahneman 2007).

During the 1990s Kahneman's interests gradually shifted from decision making and economic psychology in the direction of hedonic psychology, a field which deals with research into the factors which govern whether our life experiences are pleasant or unpleasant. We question ourselves about the nature of pleasure and pain, boredom and interest, joy and sadness, satisfaction and dissatisfaction. The research results of hedonic psychology also show the whole range of circumstances, from the biological to the societal, that cause suffering and enjoyment.

The result of these interests was the development, along with David Schkade, the concept of "focusing illusion" (Schkade and Kahneman 1998) which explains the mistakes people make when estimating the effects of different scenarios on their future happiness. The illusion occurs when people consider the impact of one specific factor on their overall happiness, they tend to greatly exaggerate the importance of that factor, while overlooking the numerous other factors that would in most cases have a greater impact. Life satisfaction, in Kahneman's opinion, is to a significant degree the result of how far we fulfill our expectations and achieve our life goals, and not, as we might mistakenly believe, due to the circumstances in which we find ourselves. The actual result of the research on happiness was also a method for collecting data on emotional well-being and time-use, called the day-reconstruction method (Kahneman et al. 2004).

At present he is studying "noise" in decision making and how to address it. Noise is an unsystematic error and it is the complement of bias, which is systematic error. The concept of noise is a useful way of thinking about decision making. One of the main reasons why simple algorithms are typically superior to human decision makers is not because of bias or systematic errors by the humans, but rather the inconsistency of human judgment (Kahneman et al. 2021).

Ever since the start of his career, Kahneman has scrupulously taken care of the methodological propriety of his research by repeating it many times. Disturbed by the quality of research conducted by social psychologists, as well as low replicability and in particular priming research, he addressed an open letter to them in 2012, which was extensively commented upon in the media. The most widely quoted of Kahneman's lines was "I see a train wreck looming …" (Kahneman 2012). Without too much exaggeration it could be said that these very words were the first shot in the methodological revolution intended to restore credibility to the results of psychological research of which we a currently witness.

Currently, he is a senior scholar and faculty member emeritus at Princeton University's Department of Psychology and Woodrow Wilson School of Public and International Affairs.

Professor Kahnemann, your research work is a stream of successes for which many of us admire you. We can read about them in many publications, which is why I'd like to start our conversation from a somewhat different angle. What has been the greatest disappointment with psychology that you have experienced in your life?

I think to some extent the replication crisis...

If so, please allow me to discuss the replication crisis a little, a subject that has been a focus of attention for several years, partially because of your famous letter in which you stated, "I see a train wreck looming." However, the trouble with replication was noticed by some a long time ago. You yourself wrote about it in your autobiography: as early as the 1960s, in order not to pollute the literature, you wanted to report only findings that you had replicated in detail at least once. Nobody was talking then about a replication crisis. How did you arrive at the conviction that the replications which nobody else really treated seriously at that time were, in fact, so important?

I think it must have been because I tried once to replicate the same result and it didn't work, so I didn't trust my results of my work unless they were replicated. I don't remember the details, but this is what must have happened. I was working in parallel on vision and problems of acuity, where all my results were absolutely replicable, and I was trying to do things in the development of psychology, and not replicating. I was replicating my work in vision, but not replicating work that I was doing on individual differences between children, I tried to do work like Walter Mischel was doing at that time, because I was very impressed with what he was doing, and I tried to do work on that, but I couldn't get the same results twice. I was probably expecting too much. That work actually led to the very first publication that I co-authored with Amos Tversky, which was basically about replication.

How many scientists do you know that have embraced standards as high as yours in regard to the replication of their research?

I didn't know anyone. It was just a decision that I made for myself, because I didn't like the idea of achieving a result and trying to write it up, and then finding out it wasn't true. But I did entirely for myself—I wasn't part of a movement or anything.

Why in your opinion did others ignore the necessity of replication for so long?

I think that by and large people did not know that they have a problem, and this was to a large extent because of a self-delusion that every investigator thought that his procedures in doing science were adequate, and didn't realize the extent of self-deception that was involved in collecting results. I think that the most basic finding was about the *p*-hacking, the extent to which investigators quite honestly take steps that cause the results to be unreliable.

In recent years there has been much said about replication projects, and indeed quite a bit is happening in that area. Some are even talking about an age of great replication in psychology. According to the PsycARTICLES database, in the last three years less than 1% of all published articles contained the word "replication" in the title. Is this a lot, or not much? How do you assess the efforts currently being made to overcome the replication crisis?

I think there really has been considerable progress. Everybody is aware of it. People are running larger samples, are more careful, are registering studies, and pre-registration is becoming standard, and even people who don't preregister think in those ways. So I think we have had a huge change, in psychology in recent years, all for the better. Many areas of psychology have changed, and it clearly is a better science than it was a decade ago because of the replication crisis.

About one year before our conversation Many Labs 2 results have been published. Half of the 28 influential psychological researches were replicated successfully. Among the latter was also the study by you and Tversky. Is 50% much or few? Is the glass half full or half empty?

Well, 50% is not enough, but some of the difficulty, I think, lies not only in the original papers, but also with the replication that they were far from perfect. Some of the failures are due to the replicating teams.

Some scientists I know take a relatively blasé attitude to the replication crisis, claiming that in other sciences, such as medicine, the percentage of successful study replications is as low or even lower. Does the fact that it can be equally bad in other sciences relieve us of the burden to care about higher rates of replicability?

I think that's ridiculous. Medicine has a problem, and they should solve that problem. We should solve ours.

What in your view should we do first to overcome the replication crisis?

I think that everything that is being done now is good. I think people should use large samples, and pre-registration of studies is a good idea. But even if you don't preregister, you are thinking as if we were. We are thinking of how we will analyze the data, and we want to plan it all ahead, so that we don't have the option of deluding ourselves about the results. This is happening now. Psychology is improving very rapidly, and I don't think the problem will go away, but it certainly is diminishing. It's been a big success.

So it's a kind of internal control system now somehow?

Yes. The culture has shifted, and even the social psychologists, who still don't believe in the replication movement, and are strongly opposed to it, have changed their procedures to get more replicable experiments. Basically, this movement has been successful.

Low reproducibility of psychologists analyses and especially social psychological research is not the only problem contemporary science is facing. For over half a century, first Jacob Cohen (1962)**, and later you and Tversky** (1971) **along with others, have pointed out that studies with low power should not be treated seriously. Has this issue been solved in contemporary studies?**

It is being solved. People are now running larger samples than ever before, and are very conscious of sample quality, so in mainstream science this problem is being solved.

In your autobiography you mentioned reading *Clinical Versus Statistical Prediction*, in which the author, Paul Meehl (1954), demonstrated the greater accuracy of actuarial forecasts based on statistical data as compared to those made by experienced clinicians making use of their experiences and intuition. When you attempted to implement those recommendations into the practice of selecting officers in the army, you ran into a wall of resistance. Even greater resistance was shown to Meehl himself and his assistants. What do you think now about that issue? Have clinicians and other specialists come around to adopting the tools offered by Meehl, significantly improved thanks to modern technology and the potentials for processing big data, or do they remain stuck in the belief that machines, never mind tables, are not capable of replacing them?

There is absolutely no question that Meehl was right, and Dawes went even further in showing that any sensible combination of variables could do as well or better than humans. So this result is now completely accepted, and I don't think there is a serious debate about the Meehl results anymore.

And now all these methods Meehl was talking about can be improved by artificial intelligence algorithms and computers.

When we started with artificial intelligence, multiple regression was better than people, then it was found that even simple models, unit weight models were better than people, and now artificial intelligence is clearly better than regression. The artificial intelligence picks up non-linear relationships in data, so that's the new standard. And by the way, the reason that people are so inferior is something that I am studying now, and this is something that we call noise and unreliability. It is because of noise more than anything else that people are inferior to simple rules. Meehl's findings were primarily a result of noise, not of bias.

Once we started talking about AI and before we move on to your current research interests please tell me what is your opinion about artificial intelligence? How important it is in our field? Because scientists are rather divided. Some of them say it will play a crucial role, and some of them say it's won't, maybe it will be useful only in improving our research methods.

I belong in the first camp. I think it's an enormously important development, and I think it's going to have huge consequences, although as I said, I am not attempting to forecast the details, but I am certainly among those who believe in it. Our model of artificial intelligence is only 10 years old, so I do not take very seriously people who say what it cannot do, because here are going to be major breakthroughs. Currently, artificial intelligence is not intelligent, so there are going to have to be big developments in that field. There is a huge amount of talent in the field at the moment, a huge amount of interest, and there is a lot of money, and so I think great things are going to happen, but I also think that some very dangerous things are going to happen. Socially, AI could have very negative consequences, but that belongs to the world of forecasting.

What kind of negative consequences do you mean?

I mean that entire professions are going to be wiped out, and the relationship between people and tools is going to change. And basically it's going to turn out that artificial drivers are better than artificial human drivers, artificial doctors are safer than real doctors, and so on. I think this is inevitable. At some stage we are going to get there, and at some stage developments in robotics are also going to make a lot of difference. It will be a very long time before you get a haircut from a robot, but I think really nobody is entirely safe from AI, and my guess is that very significant opposition is going to arise.

The future will probably show. Let's return to psychology in the meantime. In your opinion, is there any chance in the near future of the unifying theory in psychology that would reduce the proliferative character of our field a little bit? Do we need such a theory or perhaps we aren't ready for one yet?

Well, the facts speak for themselves. At the moment the field is very fragmented and there is no theory. The understanding of the brain and the mind will eventually help to organize psychology, but we are very, very far from it. So at the moment it's staying fragmented, and I'm making no predictions when it's going to change.

What are your thoughts on the views of Robyn Dawes (1994), **whom you have also met in person, and whose opinions about psychology were very radical?**

Yes, his opinions about clinical psychology were very radical. He just said that basically they were statistically uninformed, and he was obviously correct. At that time.

How are his views applied to contemporary clinical psychology?

I cannot evaluate contemporary clinical psychology. I know that there are some parts of clinical psychology, like cognitive behavior therapy, which are evidence based, and good science is being done. How much of clinical psychology has this character, I do not know.

Alongside many of your well-known achievements that can be read about in numerous places, you are also the creator (or re-creator as you mentioned in your autobiography) of a method designed to replace traditional scientific polemics, which you name adversarial collaboration. In a database of scientific articles I was able to locate just a few empirical works carried out in accordance with the recommendations of that method, and a few dozen publications involving discussion of adversarial collaboration. As the creator of that method, do you consider that it has achieved sufficient recognition among researchers, or do they rather prefer to engage in traditional polemics with one another?

Obviously it hasn't been a very successful idea. My own experiences have been interesting in this regard, and I've had several adversarial collaborations, some of them experimental, and all of those failed in the sense that we ended up with series of experiments, we understood and liked each other better, but we did not accomplish anything scientifically to clarify the field. That happened to me several times. The only two successful adversarial collaborations that I've had have been in writing papers together. I had one, that was very adversarial with people with whom I was quite angry, and who were quite angry with me (Mellers et al. 2001), I had another one with Gary Klein (Kahneman and Klein 2009), in which we became friends while carrying out that collaboration.

Do you see any chance for polemics among scholars to take on a more constructive character?

I think that this is actually likely to happen to some extent, and it's part of a new standard. The fact that people have to be much more open about their techniques and methods is going to make things much easier.

What other threads to psychology do you perceive which we haven't discussed yet?

I don't think much along those lines. Maybe I don't believe in my ability to forecast and I do not really know what is happening today. I think old people generally don't know what is happening today. What is most important is what young people going to graduate school decide what to study, and it's those decisions that make the future for the next 20 or 30 years, and I'm not following that closely.

What would you recommend for these young people trying to select a specialization at the outset of their career now?

I have no idea. I personally would go into AI, because that's a thing that I find the most thrilling, but I wouldn't give advice to anyone.

This book's subtitle is *Perspectives on Legacy Controversy and the Future of the Field*, but we have spent almost all this time discussing the controversies. Lest we upset our readers even more, let's talk about the accomplishments in our field. Which of your achievements do you feel have made the most important impact on science?

I think there is no question that prospect theory has had the most impact.

Have you made any discoveries that you yourself consider to be of importance, but which have gone unnoticed by others and not given consideration in other studies?

I think nobody ever has quite enough, and so I can think of pieces of mine that could have become more famous than they did, but I really cannot complain about anything.

Sometimes your studies are criticized for being mainly based on hypothetical questions. Critics say that the process of making a decision on how to spend or invest one's own hard-earned money is significantly different from the answers we select in a questionnaire that doesn't cost us anything. The same is true in relation to other decisions. What is your response to such criticism?

The responses could be that the work that was done in the laboratory has been extended largely through behavior economics into the market, then into the field, and it works quite well. So I think those criticisms are passé, and they aren't currently relevant. There is much successful applied work in the spirit of this research. It's successfully applied work in finance, and in social influence, and so on. I don't think that many people still talk about this area of research as disconnected from reality. This is nonsense.

In the 1990s you engaged in research on the feeling of happiness. Do you feel that these studies brought a breakthroughs of a similar caliber to those on thought processes and decision making?

My work on happiness actually was influential, but in ways that are widely known and had some influence on the design of the Gallup World Poll. The poll recognizes the distinction between life evaluation and the experience of living. The idea existed before my work, but I helped in making this acceptable.

And what is your judgment on positive psychology as a whole?

I have never been entirely sure of positive psychology. I have a temperamental opposition to a certain kind of humanism, so I haven't been very sympathetic to positive psychology. But that's entirely my character, I am a little bit cynical. There is a lot of talk about virtue and meaning, and so on, which I am usually skeptical of, but I recognize that there have been significant achievements. Some of my best friends are doing important work in that area, and in particular the work in the United Kingdom of which Richard Layard has been at the forefront is extremely useful. One of the problems that I have with the positive psychology movement is that I firmly believe that reducing misery is a more important objective for society than enhancing happiness, and this should have implications for research. That's just my own philosophical bent.

Some people call positive psychology as a psychology of feeling good. Would you agree with this label?

No, I don't think that I would use single labels. Positive psychology is many things. I don't know them all, but I think that the psychology of good feelings is not enough. It's not a friendly definition and it's not what I would choose.

What research areas are you currently focusing on? What questions are you trying to find answers to these days?

I am writing a book with two collaborators on noise in judgment and decision making. And noise is unsystematic error and it is the complement of bias, which is systematic error. Most of my career I have studied bias, and now I am very impressed by the amount of noise there is. And there are two kinds of noise. There is variability within individuals on different occasions. That's one kind of noise. And there is another kind of noise which is individual differences that exist, but shouldn't exist, like differences in judgments between judges, differences in underwriting decisions, differences in patent decisions, differences in medical decisions and so on. We are studying the variability that shouldn't exist in judgment and decision making. This is my topic today.

Many people who know your earlier books are eagerly awaiting *Noise*. I think that this will be another contribution to understanding how far off we are in our judgments about reality. Without too much exaggeration, we can say that you have dedicated nearly your entire life to showing us this. In a certain sense, you and Amos Tversky, along with your collaborators, created a new field of knowledge—about the meanderings of our minds. Did psychologists—both research and clinical—learn something from your lessons and apply that knowledge in practice?

Many people are aware of our work, and I hope its impact is positive.

What burning questions still demand answers in psychology?

I am going to disappoint you, because I really don't think that way. You know, it's not that there are questions that demand answers, it's what are the questions that are in the mind of scholars these days. I don't want to presume that I know what's going on. Clearly, there has been a trend over the last 10 or 15 years that is turning psychology to brain science. This is the biggest thing that is happening in psychology over the last several decades. Brain research is considered superior, and is taking over. That's a big development, but I have no insights about the future.

I think that a person who has participated in the most critical breakthroughs in the history of our field can afford to share his vision of a future with others. How do you imagine our field of science over the next two or three decades?

I do not believe in forecasting, because it just doesn't work and it's not the sort of exercise I engage in. I really think that the future is completely unpredictable and we have very short horizon. There are people with long term plans, and there are trends I would like to follow if I could live long enough. I am very interested in what's going to happen with artificial intelligence, more interested than in psychological thinking. The history of forecasting is not promising.

Selected Readings

Kahneman, D. (1973). *Attention and effort*. Englewood Cliffs, NJ: Prentice-Hall.

Kahneman, D. (2011). *Thinking, fast and slow*. NewYork, NY: Farrar, Straus and Giroux.

Kahneman, D., Diener, E., & Schwarz, N. (Eds.). (1999). *Well-being: The foundations of hedonic psychology*. New York, NY: Russell Sage Foundation.

Kahneman, D., Sibony, O., & Sunstein, C. R. (2021). *Noise*. London: HarperCollins.

Kahneman, D., Slovic, P., & Tversky, A. (1982). *Judgment under uncertainty: Heuristics and biases*. New York, NY: Cambridge University Press.

Kahneman, D., & Tversky, A. (Eds.). (2000). *Choices, values and frames*. New York, NY: Cambridge University Press.

References

Cohen, J. (1962). The statistical power of abnormal-social psychological research. *Journal of Abnormal and Social Psychology, 65*(3), 145–153.

Dawes, R. (1994). *House of cards*. New York, NY: Free Press.

Kahneman, D. (2007). Daniel Kahneman. In G. Lindzey & M. W. Runyan (Eds.), *A history of psychology in autobiography* (Vol. 9, pp. 115–149). Washington, DC: American Psychological Association.

Kahneman, D. (2011) *Thinking, fast and slow*. NewYork, NY: Farrar, Straus and Giroux.

Kahneman. D. (2012). The storm of doubts surrounding social priming research. *Decision Science News*. Retrieved October 17, from http://www.decisionsciencenews.com/2012/10/05/kahneman-on-the-storm-of-doubts-surrounding-social-priming-research/.

Kahneman, D., & Klein, G. (2009). Conditions for intuitive expertise: A failure to disagree. *American Psychologist, 64*(6), 515–526.

Kahneman, D., & Tversky, A. (1979). Prospect theory: An analysis of decisions under risk. *Econometrica, 47,* 313–327.

Kahneman, D., Krueger, A. B., Schkade, D. A., Schwarz, N., & Stone, A. A. (2004). A survey method for characterizing daily life experience: The day reconstruction method. *Science, 306*(5702), 1776–1780.

Kahneman, D., Sibony, O., & Sunstein, C. R. (2021). *Noise*. London: HarperCollins.

Meehl, P. E. (1954). *Clinical versus statistical prediction: A theoretical analysis and a review of the evidence*. Minneapolis, MN: University of Minnesota Press.

Mellers, A., Hertwig, R., & Kahneman, D. (2001). Do frequency representations eliminate conjunction effects? An exercise in adversarial collaboration. *Psychological Science, 12,* 269–275.

Schkade, D. A., & Kahneman, D. (1998). Does living in California make people happy? A focusing illusion in judgments of life satisfaction. *Psychological Science, 9*(5), 340–346.

Shariatmadari, D. (2015). 'What would I eliminate if I had a magic wand? Overconfidence'. *The Guardian*. Retrieved from https://www.theguardian.com/books/2015/jul/18/daniel-kahneman-books-interview.

Tversky, A., & Kahneman, D. (1971). Belief in the law of small numbers. *Psychological Bulletin, 76*(2), 105–110.

16

Carol Tavris: Writing About Psychological Science and Skepticism

Academics, like anyone else, will often turn a blind eye to evidence that they could be wrong about what they do or about a belief they hold.

Carol Tavris photo by Kathryn Jacobi

T. Witkowski, *Shaping Psychology*,
https://doi.org/10.1007/978-3-030-50003-0_16

Carol Tavris attributes the origins of her inclination to independent thinking and skepticism to her parents, Sam and Dorothy, nonreligious Jews who encouraged debate and questioning. "The thing I like best about the Jewish tradition," her secular father would say, "is that it encourages argument—including with God. Maybe especially with God!" Both of her parents were feminists and social activists. Dorothy herself became a lawyer at the age of 21, becoming a role model for Carol and teaching her the goals of feminism. Carol's parents encouraged her to ask questions about anything, from household rules to religion. They gave her books about courageous, remarkable women who were activists and pioneers in their fields, and her father taught her poetry and storytelling.

Her undergraduate studies in comparative literature and sociology—heavily influenced by Freudian theory in those days—did not satisfy her inclination for scientific inquiry. One year into her postgraduate work, she discovered a love for the scientific process and abandoned comparative literature, ultimately obtaining a Ph.D. in social psychology from the University of Michigan.

While still a doctoral student, she started working with the just-launched magazine *Psychology Today.* This experience set the direction for her future career, because she discovered there that, for her, writing about science would be much more personally satisfying than having a traditional academic career in doing science. At *Psychology Today*, she met another psychologist-editor, Carole Wade, who would become her co-author on *The Longest War: Sex Differences in Perspective* (1977/1984), which took an interdisciplinary approach to the age-old question of why gender inequality exists. In 1987, they published an introductory psychology textbook, *Psychology*, the first text to mainstream gender and culture and to feature basic principles of critical and scientific thinking. (A shorter version, *Invitation to Psychology*, followed.) That influential textbook has been in print ever since—as of 2020, through 13 editions.

Tavris's first major trade book, *Anger: The Misunderstood Emotion*, was published in 1984, and brought good psychological science to bear on many popular but unvalidated ideas—such as the notion that expressing anger reduces it, or that suppressing anger causes ulcers. In 1992, Tavris wrote *The Mismeasure of Woman*, a science-based defense of equality feminism, the view that women are neither inferior nor superior to men. The title was a tribute to Stephen Jay Gould's *The Mismeasure of Man*, as both books showed how societal prejudices can affect research. Tavris' multi-disciplinary book debunks myths about male and female brains, gender differences in abilities, PMS and other popular beliefs. In 2007, she and the world-famous social psychologist Elliot Aronson published *Mistakes Were Made (But Not by Me): Why We Justify Foolish Beliefs, Bad Decisions, and Hurtful Acts*, which explores how cognitive dissonance leads people to justify their own mistakes and harmful decisions and to hold onto beliefs and practices long after the evidence shows they should be abandoned. (The book was updated in 2015 and again in 2020.) Their colleagues in the skeptic world often enjoy referring to *Mistakes Were Made* as the "bible" of skepticism, because it explains why giving people important evidence that they might be wrong—as skeptics are forever doing—so often backfires.

Tavris's latest book, *Estrogen Matters*, written with oncologist Avrum Bluming, is another example of her determination to bring the best science to bear on complex issues of great importance to public health—even when, perhaps especially when, that science calls into question a widely held paradigm or an ideological position. In this book, the authors assess and debunk the paradigmatic belief that estrogen causes breast cancer. As their publisher describes it, the book is "a compelling defense of hormone replacement therapy, exposing the faulty science behind its fall from prominence and empowering readers to make informed decisions about their health."

Carol Tavris's articles, book reviews and op-eds have appeared in *The New York Times*, *The Wall Street Journal*, the *Los Angeles Times*, the *TLS* (formerly, the *Times Literary Supplement*), *Skeptical Inquirer* and other publications. In 2014, she began writing a column for *The Skeptic* under the heading *The Gadfly*. A Fellow of the Association for Psychological Science, she has received numerous awards for her efforts to promote science and skepticism, including an award from the Center for Inquiry's Independent Investigations Group; an honorary doctorate from Simmons College for her work in promoting critical thinking and gender equity; the Bertrand Russell Distinguished Scholar, Foundation for Critical Thinking, Sonoma State; and the Media Achievement Award from the Society for Personality and Social Psychology.

Dr. Tavris, although you haven't worked as a researcher at a university, your position in the academic world is very strong. Your name figures on a list of the 50 greatest living psychologists. Eminent scholars pay heed to your views. What is the source of your influence?

I think you exaggerate, but to the extent I have any influence, I attribute it to my efforts to persuade with evidence (and humor), to avoid dogmatism and admit changes in my thinking, and to write as clearly as I can about complex issues. You can't persuade if you write in impenetrable jargon. Though maybe you persuade some people that if they can't understand you, you must be very smart!

Several times, your publications shook not only public opinion but also some scientists, such as "Beware the Incest-Survivor Machine" (Tavris 1993), some publications on feminism (such as your Gadfly columns cautioning that a person can make a false allegation not because she or he is lying, but because of the normal mechanisms and confabulations of memory), and your recent book with Avrum Bluming (2018) on estrogen therapy. Which of these publications generated particularly strong support from the scientific community?

"Particularly" strong support? I have no idea. The psychological scientists who have done research on the false premises of recovered-memory therapy and the nature of "imagination inflation" in memory have been enormously supportive, because they know they are fighting to change beliefs that are widely held among laypersons. Social psychologists and criminologists who work to exonerate people who have been falsely convicted on the basis not just of DNA but because of faulty eyewitness testimony or coerced confessions have also supported my writing about their research. In the case of our estrogen book, we are contradicting another widely held paradigm, namely that estrogen causes breast cancer and thus is harmful to women entering menopause, when estrogen plummets sharply. We carefully assess those claims and present the surprising evidence that they are incorrect: On the contrary, women who begin taking hormone therapy at menopause have a remarkable *reduction* in the risks of heart disease, osteoporosis, and Alzheimer's. (Full disclosure:

I did not take hormones in menopause and neither Avrum nor I have taken money from any pharmaceutical company.) Many eminent physicians and medical researchers have agreed with us, but they are in the minority. For now!

You are engaged in unmasking pseudoscience in the field of psychology. Unmasking others probably doesn't make you a lot of friends, does it? How are you and your work in exposing pseudoscience perceived by other psychologists?

My life work has indeed focused on the puzzle of "why is it that when you give people good evidence that their beliefs or practices are wrong and even harmful, they don't thank you and agree?" Elliot's and my book *Mistakes were made (but not by me)* is our effort to answer that question. Over the years, I think I've come to be seen as a reliable gadfly, willing to speak truth to power, even when that "truth" or point of view is unwelcome—to feminists too. Many years ago, the feminist magazine *Ms.* had a cover story on satanic ritual abuse, that sad but devastating moral panic that was sweeping across the USA at the time. "Believe it!" the cover blared. No, thank you, I didn't, and won't.

When I hear about preposterous but popular ideas that survive in spite of the good science contradicting them, I wonder how you are able to maintain a healthy emotional distance from the injustices and harms they create. Don't they infuriate you? After all, so many of them cause real harm and human suffering.

They do infuriate and depress me. I correspond with an African American man wrongfully convicted during the daycare sex-abuse hysteria that began in the 1990s. In spite of unmistakable evidence of his innocence, the governor refuses to pardon him. All I can do is try to help him individually; support organizations that work to exonerate the innocent; and write, write, write. Sometimes I calm myself by taking the long view that human nature contains the good as well as the bad; kindness as well as cruelty; cooperation as well as competitive greed. I look at the tireless work of activists for social justice, starting with my parents, and am inspired and heartened.

Skeptics face the problem of how hard it is to get their audiences to hear and accept their arguments. How do you deal with it?

I don't expect everyone to agree with me. I certainly don't expect people to agree with me if they have entrenched, vested interests in a practice or belief that they have held for years. But I *can* hope that people who are not fully informed about an issue may be persuaded when they are given good arguments and good data. Years ago, I asked Richard McNally, the brilliant Harvard clinical psychologist who has written so powerfully about the mechanisms that lead people to believe they have been abducted by aliens (and other irrational notions), why he was willing to debate John Mack, a promoter of that very notion, in a large public forum. He told me he had no hope of changing the minds of Mack's followers—just of giving others in the audience the critical ammunition they needed to know why those beliefs were wrong. That's my goal as well.

Your attitude toward pseudoscience is atypical of the academic community. Many of them turn a blind eye to pseudoscience practices. Why, in your opinion, is this so?

I disagree with the assumption in your question. The "academic community" is very large, after all, and "pseudoscience" covers a lot of territory. I would say that most academic scholars and psychological scientists are profoundly critical of pseudoscience—if by that you mean homeopathy or astrology or any of the many marketing gimmicks that flourish in the absence of any data to support their claims. But it is true that academics, like anyone else, will be inclined to turn a blind eye to evidence that they could be wrong about what they do or about a belief they hold—especially if they have a conflict of interest, intellectual or commercial, regarding that practice or belief. That is why, in the USA, the breakdown of the former academic firewall between empirical research and its commercial application is such a danger for science. Scientists used to think it was unseemly to profit from their research; the great Jonas Salk, on being asked if he would patent his polio vaccine, said, "Can you patent the sun?" Now scientists think it's stupid not to profit from their discoveries. But when any scientific or pseudoscientific practice—such as facilitated communication or pop-psych methods of self-improvement or even well-intentioned workshops designed to eliminate prejudice and sexual harassment—is making money for its practitioners, they will almost certainly be inclined to reject, minimize or dispute disconfirming evidence.

Not only is making money from research a big problem for academics, but also publishing that research. Much research is financed from taxpayers' money, and yet if those same taxpayers want to know the research results, they have to pay for them, and not with a small fee. Access to the content of one article is a minimum of $10. The open science movement places great emphasis on open access to research results. What's your opinion on the subject?

I favor open science, if research is not self-published without peer review. It's not just the taxpayers who have to pay a lot to get the information; it's also the researchers, who often have to pay thousands of dollars out of their own pockets to get their work published. This is a disgraceful situation that is very damaging to the dissemination of scientific information.

Ignoring and even falsifying research results due to the financial interest we discussed earlier is a big problem and everything indicates that it affects even outstanding scientists. In 2019, as many as 26 of the articles by legendary psychologist Hans Eysenck (notably those on the relationship between personality, smoking and cancer) have been described as "unsafe" (King's College London 2019) **and a total of 61 his works have been submitted for retraction or correction (Marks** 2019)**. What do you think about this matter?**

I am unfamiliar with the specifics of the Eysenck retractions, so I have no idea if his publications were fraudulent, misguided, or unconsciously biased. Efforts to revisit old studies are worthy, to see which have held up and which would not pass muster today. But today's researchers face harsher pressures than I imagine was true in Eysenck's time: As I've said, when researchers are under extreme pressure to publish or perish—to publish any old thing rather than an investigation that took thought, time and effort—that increases the chance of fraud, cutting corners and manipulating statistics to make them seem more impressive than they are.

One of the very important issues you raise in your work is the gap between science and psychotherapy. One problem you have addressed concerns the many kinds of "therapies" that are not

evidence-based, often promoted by people untrained in scientific methods or even basic psychology. But even good, empirically validated forms of psychotherapy have problems: In one study (Jonsson et al. 2014), it was found that only 3% of all studies on the effectiveness of psychotherapy included monitoring of negative side effects. In another, it turned out that most of the research is conducted by researchers who do not declare a conflict of interest (Lieb et al. 2016), yet another meta-analysis of meta-analysis studies showed that only 7% of all studies contain convincing evidence confirming the effectiveness of psychotherapy (Dragioti et al. 2017). The picture that emerges from these and many other works is not very optimistic. What are your views on this subject?

Thank you for bringing me up to date on this complex issue! Since I try not to talk from ignorance, I really should shut up on this question, but it is certainly interesting and got me to thinking. It is so difficult to measure the many factors involved in psychotherapy in addition to the methods used: complexity of client's problem (spider phobia? paranoid personality disorder?), what "effectiveness" means, what "negative side effects" are, the fact that no client can be his or her own control, and, most of all, the resistance or inability of so many clients to change. The harms of psychotherapy have been noted since the 1950s, what one eminent clinical psychologist colleague of mine calls "the dirty little secret of the field." However, while I am not optimistic that people can change significantly in psychotherapy, I do think that many can benefit from the chance to speak openly with an informed and experienced listener, without feeling they are being judged. Sometimes the relief of learning that they are not crazy, "abnormal," or "sick" is therapy enough. Finding the right therapist, however, is not always easy. Especially today, given the article you told me about—Donald Meichenbaum and Scott Lilienfeld's (2018) claim that there are over 600 different therapeutic modalities currently available. Becoming an informed consumer of psychotherapy is difficult, but people who are looking for help should start by checking a therapist's credentials, experience and references. A university clinical psychology program might also be a place to start for getting referrals.

In 2018, Helen Pluckrose, James Lindsay and Peter Boghossian exposed their large-scale hoax in an attempt to ridicule what they called "grievance studies" (News at a glance 2018). In their opinion, many of the people who study minority groups that are targets of discrimination have become intellectually "corrupt," using clouds of jargon to disguise the emptiness of their ideas. Pranksters? Whistle blowers warning us against the invasion of science by ideology? Or maybe hooligans who tried to tear down important ideas? You are a feminist. What are your feelings about this incident and its consequences?

Overall, I loved what they did. Whistle-blowing pranksters. Of course the recipients of their satire would be embarrassed, upset and angry. Feminism has nothing to do with my answer; I've been as critical of unintelligible, jargony writing by feminists as by antifeminists. In fact, long ago at *Psychology Today* we reported a similar prank that had been played on leading psychology journals, which accepted a nonsensical article in jargon rather than the very same article written in clear English.

Let's talk about social psychology for a while. When you started your career working for *Psychology Today*, you witnessed a period that can fairly be called the golden era of social psychology. You interviewed Stanley Milgram (Tavris 1974), and you wrote about the most famous experiments of that era. Today, the integrity of many of them is being called into question. Gina Perry in her book *Behind The Shock Machine* questions the value of Milgram's experiments (Perry 2013). What do you think about her work?

Gina Perry discovered some important flaws in Milgram's method, notably the problems regarding debriefing and the research protocol. But I disliked her book immensely, as I explained in a review for the (London) *TLS* (Tavris 2014). To me she was doing "gotcha" journalism, trying to bring Milgram down for the sensational story of it, but without any understanding of the context of the times in which he did his work, and, for that matter, without an understanding of what social psychology is about and why so many of its findings distress people. Her tone is often snide and disparaging, attributing feelings and motives to him that reflect her own dislike of the man (and complete unawareness of what it must have been like for him, a white Jewish man, no matter how brilliant, to be at Yale, with its quotas against Jews and entrenched anti-Semitism). "Deep down," she wrote, "something about Milgram makes us uneasy"—precisely: the evidence that situations have power over our behavior. Perry insists that people's personalities and histories influence their actions, but Milgram never disputed that fact; his own experiments showed that many participants resisted. Perry tracked down one of the original subjects in the experiments, called Bill, who tried to explain to her why the studies were so valuable and why he did not regret participating, although he was one of those who went on to the end. Bill told her that people often say to him, "Nobody could ever get me to do anything like that." "Well, guess what?," he told Perry. "Yes, they can." That, of course, is the moral of the Milgram story, but Perry failed to get it. She didn't believe Bill. That's why the Milgram experiment, unlike the prison "study," remains a crucial and powerful contribution to the field.

As you mention the prison study, it is impossible to avoid discussing serious criticism raised over the Stanford Prison Experiment by Haslam and Reicher of Zimbardo's methods and claims back in 2003 and recently refreshed by Thibault Le Texier in his 2018 book and in a 2019 journal article. And Susannah Cahalan's (2019) book *The Great Pretender* casts doubt on the veracity of the experiment by David Rosenhan, years ago in the 1970s, involving sending pseudo-patients to psychiatric hospitals in the USA. This high-profile experiment commonly associated with the anti-psychiatry movement has ushered in revolutionary changes in the treatment of mental illness. What do you think about the discoveries made by these critics of "classic" studies?

The Stanford prison "experiment" was never a study and never an experiment; it was, as Leon Festinger called it at the time, a "happening." Neither was Rosenhan's work; it wasn't a scientific investigation of any kind—it was more of a gimmick to make a point that he already knew he wanted to make. (True scientists must be prepared to have their evidence *disconfirm* their hypotheses.) But Zimbardo's message was one that social psychologists supported and therefore welcomed—that the roles people are called upon to play can supersede their personal wishes and personality traits—and so, unfortunately, many set aside their discomforts with it as good psychological science. (In my intro-psych textbook with

Carole Wade, we always reported the problems with the prison study, knowing how many instructors taught it in their classes.) And Cahalan's brilliant book is long overdue. Again, at the time, many academics suppressed their discomforts with Rosenhan's methods and alleged "findings," because he, like Zimbardo, was a famous guy—at Stanford, no less. Rosenhan's claims fit the goals of the anti-psychiatry movement, as you note. But he had nothing to do with the "revolutionary changes in the treatment of mental illness" that you mention. Those had more to do with the development of anti-psychotic and antidepressant drugs than with the closing of mental institutions, which were often warehouses for suffering patients for whom there was no treatment.

I read your very positive review of Barbara Ehrenreich's (2010) ***Smile or Die*** **in which she lays down an unprecedented challenge to the foundations of positive psychology. Would you say you share similar views on positive psychology?**

Absolutely I do. I adored her sharp-witted, justified takedown of Martin Seligman's overblown promises and self-promotion, and I wrote a similar critique of positive psychology and Seligman's (2018) work in my review of his book *The Hope Circuit* for the *Wall Street Journal* (Tavris 2018).

It is indeed a ruthless assessment. However, many researchers have worked in this contemporary paradigm of positive psychology, or at least they are perceived to be its representatives. As a result of their work, a number of interesting and noteworthy concepts were created, for example, the concepts of flow by Csikszentmihalyi (1990) **or Kobasa's hardiness (Kobasa** 1979)**. Do you treat all these achievements like the work of Seligman?**

Of course not! Positive psychology has had much to contribute. I object only to the oversimplification and commodifying of many of its ideas. Flow is good, grit is good, hardiness is good—but they have exceptions and limitations; and efforts to market them to, say, improve student performance in school often turn out to be less successful than hoped. In April 2019, thanks to the efforts of psychologist James Coyne, *PLOS ONE* retracted an article about mindfulness (Gotink et al. 2015) after concluding that the authors had failed to acknowledge their commercial interests in the research, made errors of analysis, and, in Coyne's words, had written an "experimercial" pushing their institute's own products and services.

Your review contains a description of the cooperation between APA represented by Seligman and the US Army. This leads me to ask a question which many psychologists avoid answering: the implicit support for the activities of psychologists improving interrogation methods in the first decade of the twenty-first century. What is your opinion about this case?

The complicity of some American psychologists and the APA with the CIA's "war on terror," by facilitating and justifying the torture of prisoners, is the reason I resigned from the APA as soon as this report was made public.

The revelations of scholarly fraud we discussed, the absence of representativeness in psychological studies, methodological carelessness resulting in studies that are essentially non-replicable, lack of access to raw data, and other problems have led people to speak openly of a crisis in psychology as a science. Yet many scientists deny this is the case. What is your opinion—are we really in the midst of a crisis, and if we are, what are its root causes?

I think it's time to put a moratorium on the inflammatory word "crisis." I've never known a time when science in general, and psychological science in particular, has not faced serious problems—internal divisions as well as external pressures. Crises are eternal; their contents change. The criticisms of our field are important and justified, but we need a social-psychological analysis of why so many non-replicable, poorly thought out studies became so popular. They are easy to do; you can do them quickly; they don't require substantive theoretical grounding; you can do a bunch of them to beef up your CV (unfortunately, that's important for young scholars on the road to tenure, but the result is that *quantity* of publications dwarfs their *quality*); university IRB's love them because they don't seem harmful to anyone; and conflicts of interest—with funders keeping an eye on findings—are widespread. In short, *external* pressures to publish a lot of papers rather than original contributions created this latest "crisis" of imagination and method. As my dear friend and coauthor Elliot Aronson puts it, doing the original experimental work of demonstrating the mechanisms of cognitive dissonance was hard ("and fun!" he always adds), but showing that dissonance works on Thursday as well as Tuesday is easy.

As far as I know, Elliot Aronson and researchers from his generation observed behavior of the participants of their experiments. Roy Baumeister with his collaborators (2007) and Dariusz Doliński (2018) in the replication of their study showed that psychology (social psychology in particular) has become a science of self-reports and finger movements—by which I mean the shift from direct observation of behavior, widely regarded as an advance in the development of scientific methodology, to introspection. The popularity of the Mechanical Turk among scientists seems to confirm this tendency. Do you think that this shift could be the reason that some research findings work only on Thursday but not on Tuesday?

Yes.

Working on *Mistakes Were Made (But Not by Me)* had to be an extremely rewarding process for you, and also as an experience of collaboration with the legendary social psychologist Aronson. Could you tell our readers a bit about him and his views, something that we can't find in his books?

You most certainly can learn a lot about Elliot from his books, starting with his superb autobiography (Aronson 2010), *Not By Chance Alone*—the story of how he discovered social psychology, what it meant to him ("clinical psychology is about repair; social psychology is about change"), how he balanced a life of remarkable chances—being at Brandeis when Abe Maslow was there, at Stanford when Festinger was there, at Harvard when other mentors and colleagues were there—with his own wisdom and instinct to know how to take advantage of those chance opportunities. Thus, "not by chance alone" do we create our life trajectories. And the festschrift book in his honor (Gonzales et al. 2010) contains a stunning chapter by him, as well as by his many friends and students. Personally, I can tell you that he is as brilliant, witty, and wise as he comes across in his professional writing!

Let's take a closer look at psychology's achievements. In your books and articles you wrote about many of them. Which of these discoveries do you consider the most important in the development of science, and why?

You should never ask this of a textbook author, because we are at a unique vantage point to see how much is being discovered across the whole spectrum of psychological science: memory. Animal cognition. The brain. Epigenetics and chimeras. The nature of sleep. Evolutionary biology. The influence of culture on every single aspect of a human being, including what we see, how we taste, how we communicate, how we behave. Emotion. Some discoveries have been transformative: social psychology's eternal message, that we are constantly being influenced by others in our social worlds, is often unwelcome—but that doesn't make it less true, or less of a corrective on our natural hubris that "I," each individual, acts alone. Likewise the final death knell to the nature-nurture debate: we are products of both, *and* of our environments. *And* of our peer groups. *And* of chance events.

That sounds really optimistic. Which of your achievements do you consider the most important in the development of science?

None! Not being a researcher or academic, I haven't done anything to "develop science." All I can do is try to bring the best scientific information our field has to offer to the general public, students and colleagues, with a touch of skepticism and critical thinking as I go. I am most proud of my small efforts to slow the recovered-memory bandwagon, which harmed hundreds and possibly thousands of families; but Beth Loftus, Deb Poole, Steve Ceci, Maggie Bruck and many other psychological scientists were making the greatest empirical contributions. That is why *their* published contributions will live and continue to be cited, but mine were largely of and for their times.

What are the greatest challenges facing psychology in the twenty-first century?

Making sure that researchers have the academic freedom to investigate what they want and report what they find—without fear of censure, condemnation, student protest and pressures to shut up and conform.

Do you think that young people have opportunities to develop a career as an independent thinker, as you did, in the modern world?

Yes.

Is the scientific community open to people who want to work independently?

Yes and no. Yes, if the scholar writes and publishes an important, persuasive, well-researched book—such as Judith Rich Harris's (1998) brilliant *The Nurture Assumption.* No, because scientists can all too readily dismiss the ideas of someone who doesn't have the degrees and prestige that they do. Many of Harris's academic critics dismissed her arguments solely because she was an academic outsider—never mind that being an outsider was precisely what gave her the insights and perspective they lacked. She had the last laugh on them.

What advice would you give to those who decide to chart a path similar to yours?

My particular path was too idiosyncratic, too much a product of its time, to be a model for anyone else. The world always needs young people who are independent thinkers, but you can be an independent thinker in any occupation. If you are asking whether it is possible to have a career as a *self-employed* writer and scholar, then obviously the economic challenges are significant; every artist and writer needs an income. I would advise anyone interested in being a science writer to get an advanced degree and acquire a deep understanding of the methods of science, the uses and misuses of statistics, and of the particular field of science they enjoy, because with training and credentials comes greater acceptance—and a greater ability to understand and criticize new discoveries and debunk foolish ideas. We need science writers with such backgrounds more than ever, inundated as we are with studies about mental and physical health. When, in 2002, the Women's Health Initiative published its alarming "findings" that hormone therapy for women in menopause increases the risk of breast cancer, *not one science writer*, including the excellent ones at our leading newspapers, noted that the risk *was not statistically significant* and not even medically meaningful. Not one! Unfortunately, most media don't employ science writers any more, and it's very difficult to have a financially secure career solely as a writer. So the question is: what occupation will be stimulating and satisfying and allow you independence of investigation and thought? When, after getting my Ph.D., I was worried about not taking an academic job, as I had been trained to do, and instead was deciding whether to work for a *magazine*, my beloved mentor Robert Kahn said to me, "Social psychology needs good writers, too."

Selected Readings

Bluming, A., & Tavris, C. (2018). *Estrogen matters: Why taking hormones in menopause can improve women's well-being and lengthen their lives—Without raising the risk of breast cancer*. New York: Little, Brown, Spark.

Gonzales, M. H., Tavris, C., & Aronson, J. (Eds.). (2010). *The scientist and the humanist: A festschrift in honor of Elliot Aronson*. New York: Psychology Press.

Tavris, C. (1989). *Anger: The misunderstood emotion* (rev. ed.). New York: Touchstone.

Tavris, C. (1992). *The mismeasure of woman: Why women are not the better sex, the inferior sex, or the opposite sex*. New York: Touchstone.

Tavris, C. (2011). *Psychobabble and biobunk: Using psychological science to think critically about popular psychology* (3rd ed.). Upper Saddle River, NJ: Pearson.

Tavris, C., & Aronson, E. (2015/2020). *Mistakes were made (but not by me): Why we justify foolish beliefs, bad decisions, and hurtful acts* (rev. ed.). New York: Mariner Books.

Tavris, C., & Wade, C. (1984). *The longest war: Sex differences in perspective* (rev. ed.). New York: Harcourt.

Tavris, C., & Wade, C. (2001). *Psychology in perspective* (3rd ed.). Upper Saddle River, NJ: Prentice Hall.

Wade, C., Tavris, C., Sommers, S., & Shin, L. (2018). *Invitation to psychology* (7th ed.). Hoboken, NJ: Pearson Education.

Wade, C., Tavris, C., Sommers, S., & Shin, L. (2020). *Psychology* (13th ed.). Hoboken, NJ: Pearson Education.

References

Aronson, E. (2010). *Not by chance alone: My life as a social psychologist*. New York: Basic Books.

Baumeister, R. F., Vohs, K. D., & Funder, D. C. (2007). Psychology as the science of self-reports and finger movements: Whatever happened to actual behavior? *Perspectives on Psychological Science, 2*(4), 396–403.

Cahalan, S. (2019). *The great pretender: The undercover mission that changed our understanding of madness*. New York and Boston: Grand Central Publishing.

Csikszentmihalyi, M. (1990). *Flow: The psychology of optimal experience*. New York, NY: Harper & Row.

Doliński, D. (2018). Is psychology still a science of behaviour? *Social Psychological Bulletin, 13*(2). Retrieved from https://doi.org/10.5964/spb.v13i2.25025.

Dragioti, E., Karathanos, V., Gerdle, B., & Evangelou, E. (2017). Does psychotherapy work? An umbrella review of meta-analyses of randomized controlled trials. *Acta Psychiatrica Scandinavica, 136*(3), 236–246.

Ehrenreich, B. (2010). *Smile or die: How positive thinking fooled America and the world*. London: Granta Books.

Gotink, R. A., Chu, P., Busschbach, J. J. V., Benson, H., Fricchione, G. L., & Hunink, M. G. M. (2015). Standardised mindfulness-based interventions in healthcare: An overview of systematic reviews and meta-analyses of RCTs. *PLOS ONE, 10*(4). Retrieved from https://doi.org/10.1371/journal.pone.0124344.

Harris, J. R. (1998). *The nurture assumption.* London: Bloomsbury.

Haslam, S. A., & Reicher, S. (2003, Spring). Beyond Stanford: Questioning a role-based explanation of tyranny. *Society for Experimental Social Psychology Dialogue, 18,* 22–25.

Jonsson, U., Alaie, I., Parling, T., & Arnberg, F. K. (2014). Reporting of harms in randomized controlled trials of psychological interventions for mental and behavioural disorders: A review of current practice. *Contemporary Clinical Trials, 38,* 1–8.

King's College London. (2019, May). *King's College London enquiry into publications authored by Professor Hans Eysenck with Professor Ronald Grossarth-Maticek*. Retrieved from https://retractionwatch.com/wp-content/uploads/2019/10/HE-Enquiry.pdf.

Kobasa, S. C. (1979). Stressful life events, personality, and health—Inquiry into hardiness. *Journal of Personality and Social Psychology, 37*(1), 1–11.

Le Texier T. (2018). *Histoire d'un mensonge. Enquête sur l'expérience de Stanford.* Paris: Zones.

Le Texier, T. (2019, August 8). Debunking the Stanford prison experiment. *American Psychologist, 74*(7), 823–839.

Lieb, K., Osten-Sacken, J. V. D., Stoffers-Winterling, J., Reiss, N., & Barthet, J. (2016). Conflicts of interest and spin in reviews of psychological therapies: A systematic review. *BMJ Open, 6*(4). Retrieved from https://bmjopen.bmj.com/content/6/4/e010606.

Marks, D. F. (2019). The Hans Eysenck affair: Time to correct the scientific record. *Journal of Health Psychology, 24*(4), 409–420.

Meichenbaum, D., & Lilienfeld, S. O. (2018). How to spot hype in the field of psychotherapy: A 19-item checklist. *Professional Psychology: Research and Practice, 49*(1), 22–30.

News at a glance. (2018). *Science, 362*(6411), 134–136.

Perry, G. (2013). *Behind the shock machine: The untold story of the notorious Milgram psychology experiments.* New *York:* New Press.

Seligman, M. (2018). *The Hope Circuit: A psychologist's journey from helplessness to optimism.* New York, NY: Public Affairs.

Tavris, C. (1974, June). The frozen world of the familiar stranger, a conversation with Stanley Milgram. *Psychology Today, 8*(1), 70–73, 76.

Tavris, C. (1993, January 3). Beware the incest-survivor machine. *New York Times Book Review,* cover essay.

Tavris, C. (2014, July 16). "Experiments in humanity": Essay. Lessons from the Lab: Teaching Contentious Classics: Sherif, Milgram, and Harlow revisited. *Times Literary Supplement.*

Tavris, C. (2018, March 29). 'Happier?' and 'The Hope Circuit' reviews: How smiles were packaged and sold. *The Wall Street Journal.* Retrieved from https://www.wsj.com/articles/happier-and-the-hope-circuit-reviews-how-smiles-were-packaged-and-sold-1522355848.

The PLOS ONE Editors. (2019). Retraction: Standardised mindfulness-based interventions in healthcare: An overview of systematic reviews and meta-analyses of RCTs. *PLOS ONE 14*(4). Retrieved from https://doi.org/10.1371/journal.pone.0215608.

Index

T. Witkowski, *Shaping Psychology*,
https://doi.org/10.1007/978-3-030-50003-0